An Introduction
to
Quantum Chemistry

An Introduction to *Quantum Chemistry*

M. Satake
H. Yasuhisa
M.S. Sethi
Y. Mido
S. Taguchi
S.A. Iqbal

DISCOVERY PUBLISHING HOUSE
New Delhi-110 002 (INDIA)

First Published – 1996

Reprinted – 2026

ISBN: 978-81-7141-332-4

An Introduction to Quantum Chemistry

Published by:

DISCOVERY PUBLISHING HOUSE

4383/4B, Ansari Road, Darya Ganj
New Delhi-110 002 (India)
Phone: +91-11-23279245; 23253475; 43596065
Mobile: +91 9811179893 / +91 9871656464
E-mail: discoverybooksindia@gmail.com
orderdphbooks@gmail.com
namitwasan9@gmail.com
web: www.discoverypublishinggroup.com

Printed at:
Infinity Imaging Systems
Delhi

Contents

Preface

The teaching of Chemistry at the introductory stage becomes each day a more challenging task as the subject matter becomes more diverse and more complex. These challenges have evoked a series of responses—the present set of introductory chemistry monographs is one such. The teaching of chemistry recognises a number of problems that confront those who select text books. In order to overcome these problems, this volume "**An Introduction to Quantum Chemistry**"—one of about fifty in the Chmistry Monograph Series—is introduced. Each volume is independent of the others deals with one of Chemistry topics and constitutes and complete entity. Each volume is more comprehensive than can be possible in a single volume text. It is intended to provide a range of topics to cover most undergraduate and chemistry main courses of study. These volumes can be used to enrich the more conventional courses of study.

Suggestions for improvement are welcome and shall be gratefully acknowledged.

Authors

SECTION—I

MATHEMATICAL BACKGROUND

1

Introduction

Unfortunately, in spite of its great interest to chemists, quantum mechanics is not very easy to understand. It is inherently mathematical, even though definite physical ideas can always be detected behind its equations. The student might wish that he could avoid this difficulty by learning quantum mechanics without its mathematics. It is, however, impossible to acquire a satisfactory grasp of the subject without some understanding of its mathematical structure. Therefore this book does not avoid the essential mathematical details of the theory. An effort has been made to reduce the mathematical difficulties by explaining from a rather elementary point of view the features that are likely to be new to the student and by emphasizing the physical ideas expressed by the mathematics.

The physical content of quantum mechanics and its methods of "explaining" chemistry are likely to seem strange. Part of this feeling can be overcome if, in the beginning, the student understands something of the point of view of modern physical science—why the atomic theory necessarily reduced chemistry to a problem in mechanics, how scientific theories are put together, and what can and cannot be expected of them. This chapter gives a brief discussion of these points.

CHEMISTRY AS A BRANCH OF MECHANICS

Twenty five hundred years ago certain Greek philosophers suggested that matter is made up of indivisible particles, which they called atoms. According to this idea, the properties of matter result from the interactions and motions of elementary particles. The theory was highly developed in classical times.

In the past 260 years a vast number of experimental observations have been interpreted successfully in terms of the notion of the atomic constitution of matter. We have, of course, become much more sophisticated in our conceptions of atomic behavior and structure, but most of us believe in the existence of atoms almost as firmly as we believe in the existence of the chair we sit on.

Chemistry deals with the properties of matter and with their changes. According to the modern version of the atomic theory, these properties result from the motions of electrons and atomic nuclei. The motions are determined by the interaction of the elementary particles with each other's electric and magnetic fields. The principles of chemistry must therefore be a consequence of two more fundamental sciences :

(a) The science of *electrodynamics*, which deals with the behavior of electric and magnetic fields and with the forces that result when these fields interact with the elementary particles of matter.

(b) The science of *mechanics*, which deals with the actions of forces on material objects and especially with the motions that they cause.

If we seek a really penetrating understanding of chemistry, it is clear that we must begin by studying mechanics and electrodynamics. Fortunately it is possible to go a long way with a rather elementary understanding of electrodynamics. We shall therefore not have to dwell on this complicated field, which is, in fact, still not completely understood. Mechanics is quite another matter, however, and we must be prepared to make an extensive pursuit of this subject.

We have at our disposal three different systems of mechanics, each suited to limited sets of circumstances :

(1) *Classical mechanics* (that is, the mechanics of Newton) usually gives an accurate description of the motions of particles that are considerably heavier than individual atoms and that move with velocities much smaller than the velocity of light.

(2) *Quantum mechanics* is an improvement on classical mechanics that may be used in studying the motions of atomic and

subatomic particles. When applied to the motions of heavier particles it gives the same results as classical mechanics, so that classical mechanics is really an approximate form of the more general system of quantum mechanics. The limitation to particles moving slowly compared to the velocity of light continues to apply in most of quantum mechanics.

(3) The *relativistic mechanics* of Einstein is an improvement on classical mechanics adapted to heavy particles moving with velocities approaching that of light. Physicists have not yet been completely successful in formulating *relativistic quantum mechanics*, universally valid for all sizes of particles moving at any speed.

The ordinary chemical properties of matter are caused by electrons and atomic nuclei moving at velocities much less than that of light. Therefore the mechanics basis of chemistry is to be found in quantum mechanics.

THE STRUCTURE AND AIMS OF SCIENTIFIC THEORIES

The Two Components of a Scientific Theory

Scientific theories are made up of two equally important parts : first, a set of defined *concepts* in terms of which we express our experience; and second, a set of *laws* relating these concepts to one another. The importance of concepts in science often tends to be forgotten because of a confusion of the meanings of the word "law." A rule that society sets up to regulate our lives is also called a "law." Man-made laws have such a great influence on our everyday affairs that it is natural, though unjustifiable, to impose a similar meaning on the "laws of nature." We tend to think of scientific laws as rules by which nature "regulates" our existence. Just as our relationship to society is dominated by man-made laws, we assume that our relationship to nature is dominated by natural laws and we come to believe that the discovery of these laws is the chief function of science. We are likely to forget that our relationship to nature depends just as much on the concepts we use in observing nature.

In a practical sense concepts have often played a more critical role than laws in the history of scientific theories, because it is usually more difficult to recognize useful concepts than it is to find the laws relating these concepts. There are many examples of this. The laws of thermodynamics are expressed in terms of concepts of heat, energy, work, and entropy. The laws of classical mechanics use the concepts of force, velocity, and acceleration. The laws of chemistry lean heavily on a particular criterion for a chemical element. Once the concepts were formulated, however, the laws relating them usually followed without much difficulty. Modifications of laws have also sometimes become necessary, because subtle weaknesses and inconsistencies have been recognized in the definitions of certain fundamental concepts.

All students are familiar with the difficulties in understanding the concepts employed in most scientific theories, because in learning a theory they have to rediscover for themselves both the concepts and the laws that scientists have developed, and it is usually the concepts that cause them the most trouble. For instance Newton's law of motion, $f = ma$, is easy to comprehend once the concepts of force, friction, and relationship of time, position, velocity, and acceleration are understood. Similarly, in learning thermodynamics, the real battle is with the concepts of heat, work, energy, entropy, and free energy; the laws of thermodynamics are not at all difficult to understand once these concepts have been mastered.

It is interesting that quantum mechanics differs from most theories in this respect. Many of the concepts that are used in quantum mechanics are based on the familiar concepts used in classical mechanics, and since the student has usually been exposed to them earlier in his career, they do not give him any difficulty. The manner in which the laws of quantum mechanics relate these concepts to one another, however, is quite remarkable and puzzles most students. This fact seems to be one of the major difficulties that beset the student of quantum mechanics.

The Possibility of Alternative Formulations of Scientific Theories

A particular set of natural events may often be dealt with in a number of different ways, depending on the set of concepts with which one chooses to begin. Thus there are several equivalent theories of classical mechanics, each starting with a different set of fundamental variables and utilizing different laws relating these variables. In classical mechanics, Newton began with forces, masses and accelerations; Lagrange, with the so- called Lagrangian function; Hamilton, with the Hamiltonian function. In thermodynamics there are the methods of Planck, Gibbs, Lewis, Caratheodory, and Brønsted; each is based on different laws relating different sets of concepts, but in the end they all agree in their description of nature.

In quantum mechanics, also, we find that several different formulations have been put forward, based on different sets of laws but resulting in equivalent descriptions of nature. The most important of these formulations are Heisenberg's "matrix mechanics," Schroedinger's "wave mechanics," and Dirac's "symbolic method." Of these three formulations, the Schroedinger formulation is easier to understand. fortunately nearly all of the phenomena of chemistry can be satisfactorily interpreted in terms of the Schroedinger method, which is therefore very widely used in quantum chemistry.

The Schroedinger method has an advantage in that one is often able to turn to analogies with vibrating systems in order to clarify the mathematical arguments. (It is these analogies that have given the name "wave mechanics" to the Schroedinger method.) In this way one can often set up physical pictures that assist the student considerably in understanding this otherwise theory.

QUANTUM MECHANICS AS A TOOL

Some theories are used most often as tools in supplying information that is otherwise difficult to obtain. Other theories are useful primarily because they supply a framework for the logical arrangement and discussion of past observations. Still other theories are useful in both ways. The Newtonian theory of mechanics is an example of a theory that has been primarily useful as a tool in research (e.g., the location of new planets and the calculation of stresses in a rapidly accelerating component of an engine). The Darwinian theory of evolution has been more useful as a framework or language (e.g., as a basis for the classification of living things and for the interpretation of human history). Thermodynamics may be said to be useful both as a tool (e.g., the use of heats of reaction to find the temperature dependence of equilibrium constants) and as a language (e.g., the use of activity coefficients to describe the properties of solutions). In which of these respects is quantum mechanics likely to be most useful to the average chemist?

In the ten years during which quantum mechanics was first applied to chemistry, many fundamental problems were solved. The covalent bond, the periodic system of the elements, the aromatic character of benzene, the stereochemistry of carbon and other elements, the mechanism of biomolecular reactions, the existence of free radicals, van der Waals forces, the magnetic properties of matter, the conduction of electricity by metals, some of the phenomena that were finally explained in atomic terms using quantum mechanics.

The concepts introduced by quantum mechanics have given the chemist a new and powerful language with which to interpret chemical observations. This language has continued to grow in volume and usefulness. The chemist in his everyday work makes constant use of such concepts as resonance, electronegativity, hybridization, zero point energies, tunnelling through energy barriers, atomic and molecular orbitals, nonadiabatic reactions, and the classification of the states of molecules and atoms according to symmetry and angular momentum. None of these concepts can be adequately understood without a good grasp of the principles of quantum mechanics. The chemical bond itself is essentially a quantum mechanics phenomenon.

Unfortunately there are practical and theoretical limitations on the precision with which certain quantum mechanical concepts can be defined. This is true, for instance, of the resonance energy. It is important that the chemist be aware of these limitations.

2

Some Mathematical Concepts

Quantum mechanics makes use of a number of mathematical concepts which, though not familiar to the average chemist, are not difficult for him to grasp. This chapter gives a brief summary of some of these concepts. It is intended to serve as an introduction for those readers who do not already know them. The discussion in this chapter is not particularly detailed, comprehensive, or rigorous. The primary aim is to renew an acquaintance; later as these concepts are actually used.

OPERATORS

Definition and Examples

An *operator* is a symbol for a mathematical procedure that changes one function into another. For instance, the operation of taking a square root is represented by the operator, $\sqrt{\ }$. When this operator is applied to the function $x^2 + a$, we obtain a new function, $(x^2 + a)^{1/2}$. A list of some operators, along with the results of their operation on the function $x^2 + a$, is given in Table 2.1

TABLE 2.1
Some Typical Operators

Operation	Operator	Result of operation on $x^2 + a$
Multiplication by the constant, c	$c.$	$cx^2 + ca$
Taking the square	$(\ )^2$	$x^4 + 2ax^2 + a^2$
Differentiation with respect to x	d/dx	$2x$
Integration with respect to x	$\int (\ )\,dx$	$x^3/3 + ax + c$
Addition of x	$x +$	$x^2 + x + a$

Operations such as these can always be written in the form of an equation,

$$\text{(operator) (function)} = \text{(new function)}$$

For instance, $(d/dx)\,(u(x)) = du/dx$

symmetry operators form an important class which is best illustrated by means of examples. Suppose that we have a function $f(x, y, z)$, where x, y and z are the usual Cartesian coordinates of a point in space. It is helpful to think of $f(x, y, z)$ as representing some physical property that varies from point to point in space, such as the temperature in a solid body into which heat is introduced nonuniformly, or the concentration of salt in a solution that is not uniform. If the body or the solution is rotated by 180° about the z-axis, the x, y, and z axes themselves being held fixed in space, then $f(x, y, z)$ will be changed into a new function, $f'(x, y, z)$. The rotation is therefore an operation in the sense described above. (It changes one function into another.) It is generally denoted by the symbol, C_{2z}, the subscript 2 referring to the number of successive rotations needed to give a complete rotation (2 × 180° = 360°), and the subscript z referring to the axis of rotation. The symbol C_{2z} is thus an operator, and we can write

$$C_{2z}f(x, y, z) = f'(x, y, z)$$

But this operation is equivalent to replacing x in $f(x, y, z)$ by $-x$, replacing y by $-y$, and leaving z unchanged. That is,

$$C_{2z}f(x, y, z) = f(-x, -y, z)$$

Thus if

$$f(x, y, z) = (x - a)^2 + (y - b)^2 + (z - c)^2$$

then

$$C_{2z}f = (-x - a)^2 + (-y - b)^2 + (z - c)^2$$

Another symmetry operator is σ_{xy}, which represents the reflection of a body in the $x - y$ plane (i.e., the body is replaced by its image in a mirror located in the $x - y$ plane). It is easy to see that

$$\sigma_{xy}f(x, y, z) = f(x, y, -z)$$

The operation of inversion through the origin is represented by the symbol i. It means that a line is to be drawn from each point in the body through the origin of the coordinates and the point is to be moved along this line to a point equidistant from the origin on the other side of the origin.

This means that

$$if(x, y, z) = f(-x, -y, -z)$$

Problems

(1) The operator S_{nz} means rotation by 360/*n* degrees about the z-axis, followed by reflection in the *x*–*y* plane. Show that

$$S_{2z}f(x, y, z) = if(x, y, z)$$

$$S_{4z}f(x, y, z) = f(y, -x, -z)$$

(2) Show that

$$C_{6z}f(x, y, z) = f\left(\frac{1}{2}x + \frac{\sqrt{3}}{2}y, -\frac{\sqrt{3}}{2}x + \frac{1}{2}y, z\right)$$

(3) Show that if

$$f = 2x^2 + y^2 + z^2$$

then

$$C_{2y}f = C_{2z}f = C_{2x}f = if = \sigma_{xy}f = f$$

$$C_{4x}f = f$$

$$C_{4y}f = x^2 + y^2 + 2x^2$$

(4) Suppose that your right hand is placed on top of a table so that it covers the origin of the coordinates system in which *x* and *y* are in the plane of the table. Describe the results of the following operation : C_{2z}, S_{4z}, σ_{xy}, *i*. Which of these operations transforms a right hand into a left hand?

The Algebra of Operators

In a sense operators are meaningless when they stand by themselves. Yet it is possible to formulate an algebra of operators in which the operator symbols are manipulated in a manner analogous to the manipulation of number symbols in ordinary algebra. In doing this, certain simple conventions are generally followed.

Multiplication of operators : Let α, β, and γ represents three different operations. Then the expression

$$\alpha\beta\gamma f(x)$$

means that we first operate on $f(x)$ with γ, to obtain a new function, $f(x)$

$$\gamma f(x) = f(x)$$

Then $f(x)$ is operated on by β

$$\beta f(x) = f'(x)$$

and finally $f'(x)$ is operated on by α

$$\alpha f'(x) = f'(x)$$

If the same operator is applied several times in succession, it is written as a power. Thus

$$\alpha\alpha\alpha f(x) = \alpha^3 f(x)$$

The order of application of the operators is *always* from right to left as they are written. This is very important because the result of a series of operations is often different, depending on the sequence in which the operations are performed. For instance, let α represent "take the square root" and let β represent "multiply by four". Then

$$\alpha\beta f = \sqrt{4f} = 2\sqrt{f}$$

whereas

$$\beta\alpha f = 4\sqrt{f}$$

The two results are not at all the same.

If the result of two operations is the same regardless of the order in which the operations are performed, the two operators are said to *commute*. Thus the two operators just mentioned do not commute, whereas the operations, "multiply by two" and "multiply by four" do commute. The fact that many operators do not commute with one another gives operator algebra a different form from that of the ordinary algebra we learned in school.

Parentheses are often used in operator expressions in the following way. Consider the expression $(\alpha\ \beta\ f)\ (\gamma\ g)$, where α, β, and γ are operators, and f and g are functions. This expression means that the entire function $(\alpha\ \beta\ f)$ is to be multiplied (using ordinary multiplication) by the function $(\gamma\ g)$. On the other hand, in the expression $(\alpha\ \beta\ f\ \gamma\ g)$ it is to be understood that f acts as an operator that multiplies the function $\gamma\ g$, the result being operated on first by β and then by α. Thus the parentheses may be said to "insulate" operators from each other. Obviously it is always true that $(\alpha\ \beta\ f)\ (\gamma\ g) = (\gamma\ g)\ (\alpha\ \beta\ f)$, whereas it may not be true that $(\alpha\ \beta\ f\ \gamma\ g) = (\gamma\ g\ \alpha\ \beta\ f)$.

Addition and subtraction of operators : New operators can be formed by adding and subtracting operators. Thus if α and β are operators, then the new operators $\alpha + \beta$ and $\alpha - \beta$ are defined in the following way

$$(\alpha + \beta)f = af + \beta f$$

$$(\alpha - \beta)f = af - \beta f$$

Consider the function $(\alpha\beta f) - (\beta\alpha f)$, which can be represented by the expression $(\alpha\beta - \beta\alpha)f$. Then ($\alpha\beta - \beta$ *alpha*) is an operator called the *commutator* of the pair α and β. If two operators commute, their commutator is "multiply by zero". The commutator of "take the square root" and "multiply by four" is $\sqrt{4} - 4\sqrt{\ } = -2\sqrt{\ }$.

If in operating on the sum of any two functions an operator gives the same result as the sum of the operations on the two functions separately, the operator is said to be *linear*. That is, the operator α is linear if for any two functions f and g,

$$\alpha(f + g) = (\alpha f) + (\alpha g)$$

An example of a linear operator is d/dx, since

$$d/dx(f + g) = df/dx + dg/dx$$

A nonlinear operator is "take the square root," since $\sqrt{f + g} \neq \sqrt{f} + \sqrt{g}$. Linear operators have the property that they always commute with multiplication by any number

$$\alpha nf = n\alpha f \ (n = \text{a constant; } \alpha \text{ is linear})$$

Problems

(1) Is it always true that $\alpha + \beta = \beta + \alpha$ and $\alpha - \beta = -\beta + \alpha$?

(2) Let $\alpha = d/dx$, β = "multiply by four", γ = "add six", $f = x$ and $g = x^2$.

Show that

$$(\alpha\beta f)(\gamma g) = 24 + 4x^2 = (\gamma g)(\alpha\beta f)$$

$$(\alpha\beta f\gamma g) = 24 + 12x^2$$

and

$$(\gamma g\alpha\beta f) = 6 + 4x^2$$

(3) Which of the following pairs of operators commute?

(a) $\sqrt{\ }$ and $(\ )^2$

(b) d/dx and $\int_0^x (\ )dx$

(c) d/dx and $3\ +$

(d) d/dx and x.

(e) C_{4x} and σ_{xy}

(f) C_{2x} and i

(g) d/dx and C_{2x}

(h) d/dx and C_{2z}

(4) What is the commutator of x, and d/dx?

(5) Show that, if for two linear operators, α and β, it is true that $\alpha\beta - \beta\alpha = 1$, then $\alpha\beta^2 - \beta^2\alpha = 2\beta$, Extend this proof to show that $\alpha\beta^n - \beta^n\alpha = n\beta^{n-1}$.

(6) Which of the following are linear operators?

$$x\cdot,\ C_{2z},\ 6+,\ \int_0^x (\ .\ )dx,\ (\ \ )^n,\ i.$$

(7) Show that $\alpha^2 + \beta^2 = (\alpha + i\beta)(\alpha - i\beta) - i\gamma$ where γ is the commutator of α and β and $i = \sqrt{-1}$.

Eigenfunctions and Eigenvalues

If for a function, f, and an operator, α, we have

$$\alpha f = af$$

where a is a number, then f is said to be an *eigenfunction* (or *characteristic function*, or *proper function*) of the operator α, and a is said to be its *eigenvalue* (or *characteristic value* or *proper value*). For instance, e^{2x} is an eigenfunction of d/dx with an eigenvalue 2 because

$$(d/dx)e^{2x} = 2e^{2x}$$

Similarly, sin 3x is an eigenfunction of d^2/dx^2 with an eigenvalue of -9 and x^n, y^n, and z^n are eigenfunction of σ_{xy} with egenvalues 1, 1 and $(-1)^n$ respectively.

Problems

(1) Find the value of the constant a that makes e^{-ax^2} an eigenfunction of the operator $d^2/dx^2 - Bx^2$, where B is a constant. What is the corresponding eigenvalue?

(2) Show that the following are eigenfunctions of the operator in (1) if the constant a is properly chosen, and find the eigenvalues (expressed in terms of the constant B) :

(a) xe^{-ax^2}

(b) $(4ax^2 - 1)e^{-ax^2}$

(c) $(4x^3 - 3x)e^{-ax^2}$

(3) Find a function of the form $(bx^4 + cx^2 + d)\, e^{-kx^2}$ that is an eigenfunction of the operator in (1). Express the eigenvalue and the four constants, b, c, d, and k, in terms of B.

(4) Make rough graphs of the five eigenfunctions obtained in exercises (1), (2) and (3), above, all to the same scale (let $B = 1$). Make a table of the eigenvalues and note the simple relation between them. These eigenfunctions and eigenvalues are basic to the quantum mechanical theory of vibrating systems.

Differential Operators The Laplacian

Operators that involve differentiation are called differential operators. Several examples of differential operators have already been mentioned; another one that we shall have frequent occasion to use is the Laplacian operator, usually denoted by either of the symbols ∇^2 (called "del squared" or "nabla squared") or Δ. It is defined as

$$\nabla^2 \equiv \Delta \equiv \frac{\partial^2}{\partial x^2} + \frac{\partial^2}{\partial y^2} + \frac{\partial^2}{\partial z^2} \tag{2.1}$$

where x, y, and z are Cartesian coordinates. It is called the Laplacian operator from its occurrence in Laplace's differential equation

$$\frac{\partial^2 f}{\partial x^2} + \frac{\partial^2 f}{\partial y^2} + \frac{\partial^2 f}{\partial z^2} = 0$$

or

$$\nabla^2 f = 0$$

If one transforms the Cartesian coordinates into polar coordinates, (r, θ, φ), setting (see Fig. 2.1)

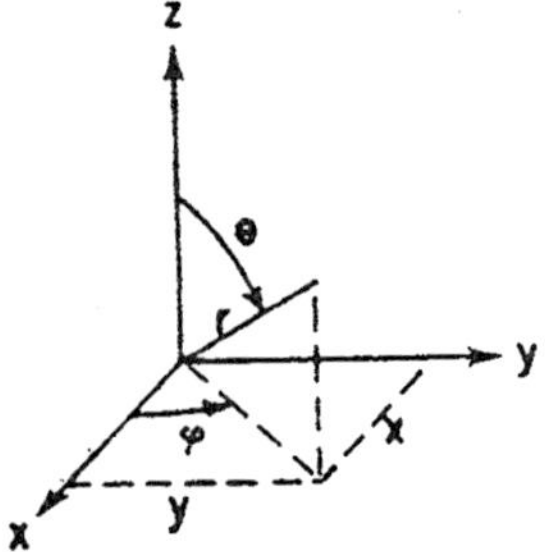

Fig. 2.1 Relation of Cartesian coordinates to polar coordinates.

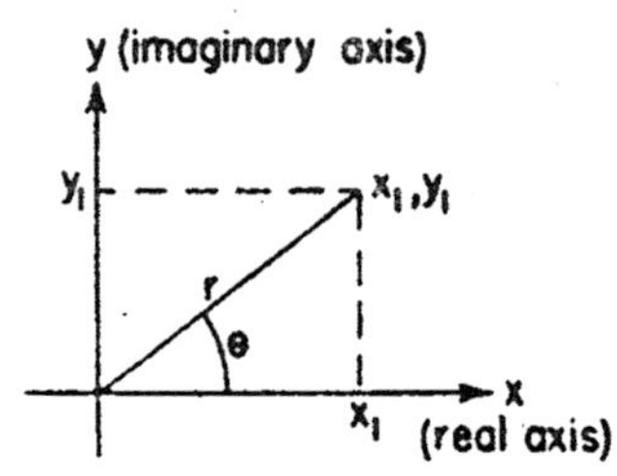

Fig. 2.2. The Argand diagram.

$$x = r \sin\theta \cos\varphi \qquad\qquad r = \sqrt{x^2 + y^2 + z^2}$$

$$x = r \sin\theta \cos\varphi \qquad \text{or} \qquad \theta = \cos^{-1}(z/r)$$

$$z = r \cos\theta \qquad\qquad \varphi = \tan^{-1}(y/x)$$

the Laplacian operator takes the form

$$\begin{aligned}\nabla^2 &= \frac{1}{r^2}\frac{\partial}{\partial r}\left(r^2 \frac{\partial}{\partial r}\right) + \frac{1}{r^2 \sin\theta}\frac{\partial}{\partial\theta}\left(\sin\theta \frac{\partial}{\partial\theta}\right) + \frac{1}{r^2 \sin^2\theta}\frac{\partial^2}{\partial\varphi^2} \\ &= \frac{\partial^2}{\partial r^2} + \frac{2}{r}\frac{\partial}{\partial r} + \frac{1}{r^2}\frac{\partial^2}{\partial\theta^2} + \frac{\cot\theta}{r^2}\frac{\partial}{\partial\theta} + \frac{1}{r^2\sin^2\theta}\frac{\partial^2}{\partial\varphi^2}\end{aligned} \tag{2.2}$$

Problems

Show that the following are eigenfunctions of the Laplacian and find the eigenvalues :

(a) $\sin kx \sin my \sin nz$ (b) $1/r$ (c) $\frac{1}{r}\sin kr$

(d) $\left[\frac{\sin kr}{k^2r^2} - \frac{\cos kr}{kr}\right]\cos\theta$ (e) $\frac{1}{r^2}\cos\theta$ (f) $\frac{1}{r^2}\sin\theta\sin\varphi$

(g) $\frac{1}{r^3}\sin^2\theta\sin 2\varphi$ (h) $\frac{1}{r^3}\sin\theta\cos\theta\cos\varphi$

COMPLEX NUMBERS

The student will recall that the quantity $\sqrt{-1}$ is known as the *imaginary quantity* and is ordinarily denoted by the symbol i or j. It gives rise to the so-called complex numbers, that is, numbers that can be written as

$$z = x + iy$$

x and y being ordinary real numbers. The *conjugate complex* of the complex number z is simply the number with a minus sign written in front of the i; it is represented by the symbol z^* or $\bar{z}$. Thus if

$$z = x + iy$$

then

$$z^* \equiv \bar{z} \equiv x - iy$$

It is important to note that

$$zz^* = x^2 + y^2$$

which is always a positive number. The quantity

$$|z| = \sqrt{zz^*}$$

is known as the *absolute value of* z; it is always positive and real.

Our comprehension of the bahavior of real numbers is immensely aided by looking at them from a geometrical point of view. If we draw a straight line of infinite length, mark off equal increments of length and then proceed to number each mark, starting at $-\infty$, ..., -2,

-1, o, 1, 2,..., and ending at $+\infty$, then it is possible to locate every real number on this line. If we try to do the same thing for complex numbers we are faced with the difficulty that every complex number has two parts—a real part, x in the expression $z = x + iy$, and an imaginary part, y. This means that we need two degrees of freedom to represent a complex number, and it is natural to choose a plane for this purpose. One draws two perpendicular axes on a piece of paper, labelling the horizontal axis the x-axis or real axis, and the vertical one the y-axis, or imaginary axis (Fig. 2.2). Then every conceivable complex number is represented by a point in the x-y plane. The number $z_1 = x_1 + iy_1$, for instance, corresponds to the point (x_1, y_1) in Fig. 2.2. Such a means of representing complex numbers is called an *Argand diagram.*

It is not necessary, however, to use Cartesian coordinates, x, y, in such a representation; it is also possible to represent a complex number by its polar coordinates, r, θ. Thus,

$$z = x + iy = r \cos \theta + ir \sin \theta = r(\cos \theta + i \sin \theta)$$

Now it happens that

$$\cos \theta + i \sin \theta = e^{i\theta} \tag{2.3}$$

a very important relation which is proven below. Thus in the polar coordinate representation we have

$$z = re^{i\theta} \tag{2.4}$$

It is also easy to see that $\quad z^* = re^{-i\theta}$

and $\quad zz^* = r^2e^{i\theta - i\theta} = r^2$

Thus $\quad r = |z| \quad (2.5)$

The quantity r is therefore simply the absolute value of the complex number; it is also sometimes called the *modulus* of z. The angle θ is sometimes called the *amplitude* or *phase* of z.

Derivation of equation (2.3). It is well known that the following series are valid for all values of x

$$e^x = 1 + x + (x^2/2!) + (x^3/3!) + (x^4/4!) + \ldots$$

$$\sin x = x - (x^3/3!) + (x^5/5!) - \ldots$$

$$\cos x = 1 - (x^2/2!) + (x^4/4!) - \ldots$$

If we write x = iθ and remember that $i^2 = -1$, $i^3 = \times\, i$, $i^4 = 1$, etc., then we find

$$e^{i\theta} = 1 + i\theta - \theta^2/2! - i\theta^3/3! + \theta^4/3! + \theta^4/4! + i\theta^5/5! - \theta^6/6! - \ldots$$

$$= (1 - \theta^2/2! + \theta^4/4! - \ldots) + i(\theta - \theta^3/3! + \theta^5/5! - \ldots)$$

$$= \cos\theta + i\sin\theta$$

Problems

(1) Show that

$$\sin x = \frac{e^{ix} - e^{-ix}}{2i} \text{ and } \cos x = \frac{e^{ix} + e^{-ix}}{2}$$

(2) Using the result of (1), show that if *m* and *n* are integers,

$$\int_{-\pi}^{\pi} \cos mx \cos nx \, dx = \int_{-\pi}^{\pi} \sin mx \sin nx \, dx = \begin{cases} 0 \text{ if } m \neq n \\ \pi \text{ if } m = n \neq 0 \end{cases}$$

$$\int_{-\pi}^{\pi} \cos mx \sin nx \, dx = 0$$

for all integral values of m and n.

(3) Show that (a) $(2 + 3i)/(1 + i) = \frac{5}{2} + \frac{1}{2}i$

(b) $\sqrt{1 + i} = 2^{1/4} (\cos \pi/8 + i \sin \pi/8)$

(c) $\log (x + iy) = \log |x + iy| + i \tan^{-1}(y/x)$

(d) $\sin (x + iy) = \sin x \cosh y + i \cos x \sinh y$

Note : $\cosh y$ = "hyperbolic cosine of y" $= \frac{1}{2}(e^y + e^{-y})$

$\sinh y$ = "hyperbolic sine of y" $= \frac{1}{2}(e^y - e^{-y})$

(4) Give a simple rule for finding the location of the complex conjugate of a complex number on the Argand diagram.

WELL BEHAVED FUNCTIONS

A function $\psi(x, y,...)$ is said to be well behaved if three conditions are satisfied :

(1) It is *single valued* (i.e., for each value of the variables $x, y,...$; there is only one value of the function ψ). This means that the function must not "bend over" on itself, as $f(x)$ does in the vicinity of x' in Fig. 2.3.

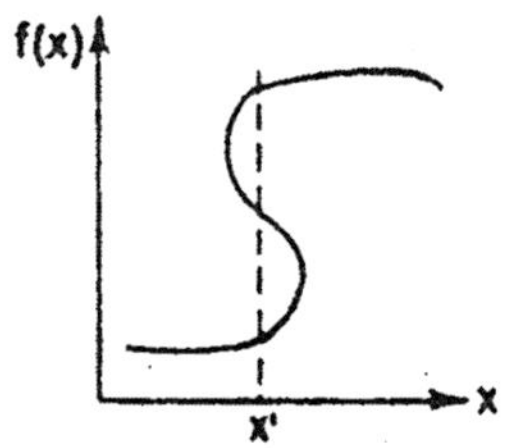

Fig. 2.3. A function of x that is not single valued at $x = x'$.

Fig. 2.4. A function of x that is not continuous at $x = x''$.

(2) It is *continuous* (i.e., there are no sudden changes in ψ as its variables are changed. Thus Fig. 2.4 shows a function $g(x)$ that is discontinuous at x'').

(3) *The integral of the square of the absolute value of the function is finite*, the limits of integration being the range of all of the variables. That is, if x can vary from a to a', y can vary from b to b', etc., then $\int_a^{a'} \int_b^{b'} ...\psi\psi^* \, dx \, dy... =$ a finite number. An example of a well behaved function is $e^{-x^2/2}$, where x can have values between $-\infty$ and $+\infty$, because $\int_{-\infty}^{\infty} e^{-x^2} \, dx = \sqrt{\pi}$. On the other hand, the function x^2 is not well behaved in this range of values of x because $\int_{-\infty}^{\infty} x^4 \, dx$ is infinite. Sometimes we are concerned with variables which can have only a limited range of values. For instance θ in $e^{i\theta}$ can vary only from 0 to 2π if it represents a polar coordinate. Thus $e^{i\theta}$ is well behaved since $\int_0^{2\pi} e^{i\theta} \, d\theta = 2\pi$. On the other hand, e^x and $e^{i\theta}$ are not well behaved if x can vary from $-\infty$ to $+\infty$.

VECTORS

Quantities which have only magnitude are known as *scalars* (for instance, mass, volume, time, wave length, temperature, entropy). Quantities which have a direction in space as well as magnitude are called *vectors* (e.g., force, velocity, concentration gradient, flow of heat through a body). The symbols for scalars and vectors are usually distinguished by underlining the vectors (or in printed books, by writing them in bold faced type). Vectors can be represented by arrows whose length is equal to their magnitude and whose direction is parallel to the direction of the vector.

Addition of Vectors

Two vectors can be added together to give a new vector. This is done by the so-called parallelogram law: the sum of two vectors, $\boldsymbol{A}$ and $\boldsymbol{B}$, is the diagonal of the parallelogram of which $\boldsymbol{A}$ and $\boldsymbol{B}$ are the adjacent sides (Fig. 2.5).

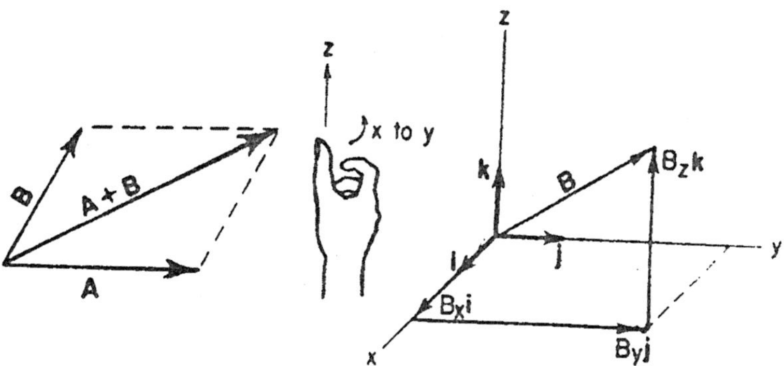

Fig. 2.5. The addition of vectors.

Fig. 2.6. The resolution of a vector into components along the *x*, *y*, and *z* axes.

The vector $-\boldsymbol{B}$ is simply the vector $\boldsymbol{B}$ turned about in space by 180°. Obviously the sum of $\boldsymbol{B}$ and $-\boldsymbol{B}$ is a vector of zero length, or

$$\boldsymbol{B} + (-\boldsymbol{B}) = 0$$

A *unit vector* is one whose magnitude is unity. Any vector can be expressed as the product of its magnitude (a scalar generally represented by placing vertical lines on either side of the vector) and a unit vector in the direction of the vector. Thus if $\boldsymbol{b}$ is a unit vector in the direction of $\boldsymbol{B}$, we can write

$$\boldsymbol{B} = |\boldsymbol{B}|\ \boldsymbol{b}$$

Resolution of Vectors into Components

Just as any group of vectors can be added together to give a single vector, so any one vector can be considered to be the sum of a group of other vectors, each having a specified direction. The latter are called *components* of the given vector in these directions. The most common system of resolution of a vector into components is the following, and when one speaks of "the components" of a vector, it is usually this system which is implied.

A Cartesian coordinate system is set up as shown in Fig. 2.6, with unit vectors, $\boldsymbol{i}$, $\boldsymbol{j}$, and $\boldsymbol{k}$ directed respectively along the positive x, y, and z axes. Unless otherwise stated, it is always understood that x, y, and z form a *right-handed system;* that is, if y and z in Fig. 2.6 are in the plane of the paper, x sticks *out of* the paper toward the reader; or, if one grasps the z-axis with one's right hand, with fingers pointing from positive x toward positive y, then the thumb points in the positive z directions.

Any vector can be resolved into the sum of three vectors in the $\boldsymbol{i}$, $\boldsymbol{j}$, and $\boldsymbol{k}$ directions. Thus if (Fig. 2.6)

$$\boldsymbol{B} = B_x\boldsymbol{i} + B_y\boldsymbol{j} + B_z\boldsymbol{k}$$

then the quantities B_z, B_y, and B_z completely specify the magnitude and direction of $\boldsymbol{B}$. They are the components of $\boldsymbol{B}$. The manipulation of vectors by means of their components is very convenient for many purposes. For instance, if vectors $\boldsymbol{A}$ and $\boldsymbol{B}$ have components (A_x, A_y, A_z) and (B_x, B_y, B_z), then it is easy to see that

$$\boldsymbol{A} + \boldsymbol{B} = (A_x + B_z)\boldsymbol{i} + (A_y + B_y)\boldsymbol{j} + (A_z + B_z)\boldsymbol{k}$$

Thus the components of the sum of two vectors are simply the sums of the corresponding pairs of components. It is much easier and more accurate to add vectors in this way than to add them by drawing parallelograms.

Problems

A cube has sides of length 2. A right- handed Cartesian coordinate system is set up with its origin at the center of the cube and with axes parallel to the sides of the cube (see Fig. 2.7). Write down the following vectors in terms of the components parallel to $\boldsymbol{i}, \boldsymbol{j}$, and $\boldsymbol{k}$: $\boldsymbol{OA}$, $\boldsymbol{AB}$, $\boldsymbol{OB}$, $\boldsymbol{OC}$. Show that $\boldsymbol{OB} = \boldsymbol{OA} + \boldsymbol{AB}$, both according to the parallelogram law and by the addition of components.

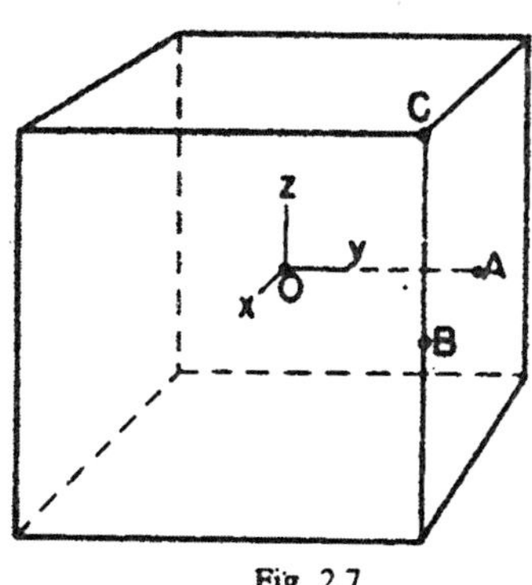

Fig. 2.7.

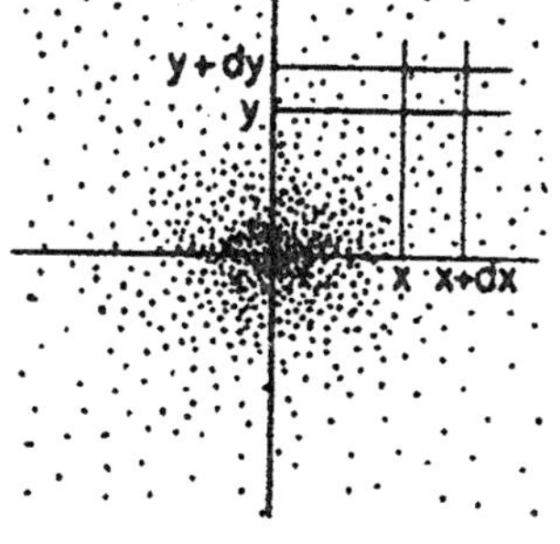

Fig. 2.8. A two-dimensional probability distribution resulting from shooting at a target.

Multiplication of Vectors

There are two different methods of combining vectors by multiplication. They are known as the scalar product and the vector product.

The scalar product of two vectors, $\boldsymbol{A}$ and $\boldsymbol{B}$, is a scalar quantity defined by

$$\boldsymbol{A} \cdot \boldsymbol{B} = |\boldsymbol{A}|\ |\boldsymbol{B}| \cos \theta \tag{2.6}$$

where θ is the angle between the two vectors. The scalar product, $\boldsymbol{A} \cdot \boldsymbol{B}$, is also sometimes written as $(\boldsymbol{A}, \boldsymbol{B})$. It is evidently equal to the length of one of the two vectors ($|\boldsymbol{A}|$) multiplied by the projection of the other vector on it ($|\boldsymbol{B}| \cos \theta$). An important example is the work done in moving against a force. Displacement and force are both vectors, and work is a scalar defined by the product of the displacement times the component of the force in the direction of the displacement. If $\boldsymbol{F}$ is the force, $\boldsymbol{s}$ is the displacement, and θ is the angle between $\boldsymbol{F}$ and $\boldsymbol{s}$, then it is clear that the work done in moving $\boldsymbol{F}$ a distance $\boldsymbol{s}$ is

$$\text{Work} = (|\boldsymbol{F}| \cos \theta)\, |\boldsymbol{s}| = \boldsymbol{F}\cdot\boldsymbol{s}$$

It is easy to see that

$$\boldsymbol{i}\cdot\boldsymbol{i} = \boldsymbol{j}\cdot\boldsymbol{j} = \boldsymbol{k}\cdot\boldsymbol{k} = 1$$

$$\boldsymbol{i}\cdot\boldsymbol{j} = \boldsymbol{j}\cdot\boldsymbol{k} = \boldsymbol{k}\cdot\boldsymbol{j} = \boldsymbol{k}\cdot\boldsymbol{i} = \boldsymbol{i}\cdot\boldsymbol{k} = 0$$

Furthermore $\quad \boldsymbol{A}\boldsymbol{B} = (A_x\boldsymbol{i} + A_y\boldsymbol{j} + A_z\boldsymbol{k}) \cdot (B_x\boldsymbol{i} + B_y\boldsymbol{j} + B_z\boldsymbol{k})$

$$= A_xB_x + A_yB_y + A_zB_z \tag{2.7}$$

Problems

(1) Prove that $\boldsymbol{A}\cdot\boldsymbol{A} = |\boldsymbol{A}|^2$ for any vector.

(2) Prove that $\boldsymbol{A}\cdot\boldsymbol{B} = \boldsymbol{B}\cdot\boldsymbol{A}$; that is, that scalar multiplication is commutative.

(3) Find the angle between two diagonals passing from one corner to another through the center of cube, using the relation

$$\cos \theta = \boldsymbol{A}\cdot\boldsymbol{B}/(|\boldsymbol{A}|\, |\boldsymbol{B}|)$$

where $\boldsymbol{A}$ and $\boldsymbol{B}$ are the two diagonals in question. This angle is the tetrahedral angle between C–H bonds in methane, etc., which is so important in the stereochemistry of carbon.

The *vector product* of $\boldsymbol{A}$ and $\boldsymbol{B}$ is the vector defined by

$$\boldsymbol{A} \times \boldsymbol{B} = |\boldsymbol{A}|\, |\boldsymbol{B}|\, \boldsymbol{c} \sin \theta \tag{2.8}$$

where θ is the angle between $\boldsymbol{A}$ and $\boldsymbol{B}$ and $\boldsymbol{c}$ is a unit vector perpendicular to $\boldsymbol{A}$ and $\boldsymbol{B}$ such that $\boldsymbol{A}$, $\boldsymbol{B}$, $\boldsymbol{c}$ forms a right- handed system. (If the fingers of the right hand point from $\boldsymbol{A}$ to $\boldsymbol{B}$, the thumb points along $\boldsymbol{c}$.) The vector product, $\boldsymbol{A} \times \boldsymbol{B}$, is also sometimes written $[\boldsymbol{A}, \boldsymbol{B}]$. The length of $\boldsymbol{A} \times \boldsymbol{B}$ is numerically equal to the area of the parallelogram whose sides are $\boldsymbol{A}$ and $\boldsymbol{B}$, and its direction is perpendicular to the plane of this parallelogram. It is easy to see that

$$\begin{array}{ll} \boldsymbol{i} \times \boldsymbol{i} = \boldsymbol{j} \times \boldsymbol{j} = & \boldsymbol{k} \times \boldsymbol{k} = 0 \\ \boldsymbol{i} \times \boldsymbol{j} = \boldsymbol{k} & \boldsymbol{j} \times \boldsymbol{i} = -\boldsymbol{k} \\ \boldsymbol{j} \times \boldsymbol{k} = \boldsymbol{i} & \boldsymbol{k} \times \boldsymbol{j} = -\boldsymbol{i} \\ \boldsymbol{k} \times \boldsymbol{i} = \boldsymbol{j} & \boldsymbol{i} \times \boldsymbol{k} = -\boldsymbol{j} \end{array} \tag{2.9}$$

Furthermore, it is always true that

$$\boldsymbol{A} \times \boldsymbol{B} = -\boldsymbol{B} \times \boldsymbol{A}$$

That is, vector multiplication is not commutative.

Also

$$\begin{aligned}
\boldsymbol{A} \times \boldsymbol{B} &= (A_x\boldsymbol{i} + A_y\boldsymbol{j} + A_z\boldsymbol{k}) \times (B_x\boldsymbol{i} + B_y\boldsymbol{j} + B_z\boldsymbol{k}) \\
&= A_xB_x\boldsymbol{i} \times \boldsymbol{i} + A_xB_y\boldsymbol{i} \times \boldsymbol{j} + A_xB_z\boldsymbol{i} \times \boldsymbol{k} \\
&+ A_yB_x\boldsymbol{j} \times \boldsymbol{i} + A_yB_y\boldsymbol{j} \times \boldsymbol{j} + A_yB_z\boldsymbol{j} \times \boldsymbol{k} \qquad (2.10) \\
&+ A_zB_x\boldsymbol{k} \times \boldsymbol{i} + A_zB_y\boldsymbol{k} \times \boldsymbol{j} + A_zB_z\boldsymbol{k} \times \boldsymbol{k} \\
&= (A_yB_z - A_zB_y)\boldsymbol{i} + (A_zB_x - A_xB_z)\boldsymbol{j} + (A_xB_y - A_yB_x)\boldsymbol{k}
\end{aligned}$$

This can also be written in the form of the determinant

$$\boldsymbol{A} \times \boldsymbol{B} = \begin{vmatrix} \boldsymbol{i} & \boldsymbol{j} & \boldsymbol{k} \\ A_x & A_y & A_z \\ B_x & B_y & B_z \end{vmatrix} \qquad (2.11)$$

An example of the occurrence of the vector product that is of considerable importance in quantum mechanics is the angular momentum of a particle moving about a fixed point. Let the mass of the particle be m and suppose that it moves with a velocity $\boldsymbol{v}$ (a vector) at a distance $\boldsymbol{r}$ (also a vector) from the point. Then the angular momentum $\boldsymbol{M}$ is defined as the vector whose direction is perpendicular to both $\boldsymbol{v}$ and $\boldsymbol{r}$, and whose sense is such that $\boldsymbol{r}$ $\boldsymbol{v}$, $\boldsymbol{M}$ form a right-handed system. Its magnitude is the product of the component of linear momentum perpendicular to $\boldsymbol{r}$ with the distance from the point. Thus

$$|\boldsymbol{M}| = m|\boldsymbol{v}|\,|\boldsymbol{r}| \sin\theta$$

It is evident that the angular momentum is simply the vector product of the linear momentum and the radius vector, or

$$\boldsymbol{M} = m\boldsymbol{r} \times \boldsymbol{v} \qquad (2.12)$$

PROBABILITY FUNCTIONS AND AVERAGE VALUES

If one shoots a gun at a target many times a pattern such as that shown in Fig. 2.8 can be expected. Let N be the number of shots hitting the target. The location of a point on the target can be specified by giving its coordinates (x, y) in a suitable two dimensional Cartesian coordinate system. Then if N is large and the marksmanship is random it is possible to set up a function, $f(x, y)$, such that

$$Nf(x, y)\,dx\,dy$$

is the number of hits in the rectangular region of the target between x, $x + dx$, y, and $y + dy$. The function $f(x, y)$ is called a *probability distribution* (or *probability function*).

It is obvious that since the total number of shots hitting someplace on the target is N, then

$$\iint_{\text{target}} Nf\,dx\,dy = N$$

or

$$\iint_{\text{target}} f\,dx\,dy = 1 \tag{2.13}$$

The function f defined in this way is said to be *normalized.*

Suppose that a prize of $g(x, y)$ dollars is given for each shot hitting the point (x, y). Then the total prize won with N shots will be

$$G = \iint Ngf\,dx\,dy$$

the integration being performed over the entire target. The average prize per shot will be

$$\bar{g} = \frac{G}{N} = \frac{\iint gf\,dx\,dy}{\iint f\,dx\,dy} \tag{2.14}$$

or if f is normalized

$$\bar{g} = \iint gf\,dx\,dy \tag{2.15}$$

It is easy to generalize this expression so as to be able to calculate the average value of any property associated with a system for which a probability function can be written.

Problems

(1) Suppose that the probability function for hitting the point (x, y) on a target is $f = A\exp(-(x^2 + y^2)/\sigma^2)$ and that a prize awarded is

$$g = \$2.000 \text{ if } \sqrt{x^2 + y^2} < a$$
$$g = 0 \text{ if } \sqrt{x^2 + y^2} > a$$

The constant σ is a measure of the accuracy of the marksmanship of the shooter, a small value indicating good markmanship. Find the value of the constant A which normalizes the function f. If a charge of one dollar is made for each shot fired, what is the largest value of σ that the marksman can have if he is not to lose money in the shooting match? (Answer : $\sigma = 1.20\ a$).

(2) The probability function for a molecular velocity c in an ideal gas at temperature T is

$$f(c) = Ac^2 \exp(-mc^2/2kT)$$

where m is the molecular mass, k is the Boltzmann gas constant, and A is the normalization constant. If the kinetic energy of a molecule moving with velocity c is $\frac{1}{2}mc^2$, find the mean kinetic energy per molecule in the gas. Note that

$$\int_0^\infty x^{2n} e^{-ax^2} dx = [(2n)!/(2^{2n+1} n!a^n)] \sqrt{\pi/a}$$

SERIES EXPANSION OF FUNCTIONS

Certain sets of functions have the remarkable property that they may be used as the basis for the expansion of other arbitrary functions. Such sets of functions play a very important part in quantum mechanics, and in order to familiarize the student with the notion of series expansions we shall consider two special types of series.

Power Series

The student is undoubtedly already familiar with the Taylor-Maclaurin series, which is based on the integral powers $(x - x_0)$, where x_0 is a constant. In this series we have

$$f(x) = \sum_{n=0}^{\infty} a_n (x - x_0)^n$$

where the constant coefficients a_n are given by

$$a_n = (1/n!)\,(d^nf/dx^n)\,x = x_0$$

Such a series is often valid over only a limited range of values of x. For instance, the expansion of the function $1/(1 - x)$ about $x_0 = 0$ gives

$$1/(1 - x) = 1 + x + x^2 + x^3 + \dots$$

which is valid only in the range $-1 < x < 1$, since outside of this range the series fails to converge. The expansions

$$e^x = 1 + x + (x^2/2!) + (x^3/3!) + \dots$$

$$\sin x = x - (x^3/3!) + (x^5/5!) - \dots$$

$$\cos x = 1 - (x^2/2!) + (x^4/4!) - \dots$$

are, on the other hand, valid for all values of x. Many functions cannot be expanded at all about certain values of x_0, because some or all of their derivatives are infinite. An example is $f(x) = x^{1/2}$, which cannot be expanded about $x_0 = 0$.

Fourier Series

A more powerful basic set of expansion functions was unearthed in 1807 by Fourier, who showed that in the range $-\pi < x < \pi$ almost any function, $f(x)$, can be expressed in the form

$$f(x) = \sum_{n=0}^{\infty} (a_n \cos nx + b_n \sin nx) \qquad (2.16)$$

where a_n and b_n are constant whose values will be given below. A function can always be expanded in terms of sines and cosines in this way if it satisfies the following conditions : (1) If it has only a finite number of discontinuities. (2) If it is single valued except at the discontinuities. (3) If it oscillates only a finite number of times.

The Fourier series differs from the power series in several important respects. The range of convergence of the power series is different for different functions; for the Fourier series (2.16) on the other hand, the range of convergence is always $-\pi$ to $+\pi$. Furthermore, the power series expansion is not possible at all for many unremarkable functions, whereas the Fourier series will represent all but the most extraordinary functions.

The validity of the Fourier expansion is by no means self- evident. It is interesting that in the half-century before Fourier made his discovery, several mathematicians (notably Lagrange and Euler) had encountered these expansions in specific problems, but had failed to generalize them because they seemed intuitively unreasonable. Fourier encountered some difficulty in having his theory accepted for the same reason.

The constant a_n and b_n in equation (2.16) are readily evaluated because of the fact that,

$$\int_{-\pi}^{\pi} \cos mx \sin nx \, dx = 0 \tag{2.17}$$

m and n being integers, and

$$\left.\begin{aligned} &\int_{-\pi}^{\pi} \cos mx \cos nx \, dx = 0 \qquad (2.18) \\ &\int_{-\pi}^{\pi} \sin mx \sin nx \, dx = 0 \qquad (2.19) \end{aligned}\right\} (mx \neq n)$$

If the integral of the product of two functions vanishes in this way the functions are said to be *orthogonal.* (Of course, the orthogonality of a pair of functions depends on the limits of the integration, and these must be stated if they are not obvious from context.) If m and n in equations (2.18) and (2.19) are equal, we have for $m \neq 0$

$$\int_{-\pi}^{\pi} \sin^2 mx \, dx = \int_{-\pi}^{\pi} \cos^2 mx \, dx = \pi \tag{2.20}$$

Therefore, if we multiply equation (2.16) on both sides by cos kx and integrate from $-\pi$ to π, we find that all terms on the right other than the kth term vanish, giving

$$a_k = (1/\pi) \; f(x) \cos kx \, dx \tag{2.21}$$

The coefficient a_0 is, however, a special case, and we find

$$a_0 = (1/2\pi) \int_{-\pi}^{\pi} (fx) \, dx \tag{2.22}$$

Similarly, multiplying by sin kx, we find

$$b_k = (1/\pi) \int_{-\pi}^{\pi} f(x) \sin kx \, dx \tag{2.23}$$

Let us expand the function $f(x) = x$ as an illustration of the Fourier series. Then

$$a_k = (1/\pi) \int_{-\pi}^{\pi} x \cos kx \, dx = 0$$

as is easily shown by making a graph of the integrand, from which it will be seen that the area under the curve to the left of $x = 0$ is equal and opposite in sign to that on the right, so that the two areas cancel.*

$$b_k = (1/\pi) \int_{-\pi}^{\pi} x \sin kx \, dx = (1/\pi k^2) \int_{-\pi k}^{\pi k} y \sin y \, dy$$

where $y = kx$. Integrating by parts

$$b_k = (1/\pi k^2) \left[-y \cos y + \int \cos y \, dy \right]_{-k\pi}^{k\pi} = (-1)^{k-1} (2/k)$$

Thus

$$x \; 2 \sin x - \sin 2x + (2/3) \sin 3x - (1/2) \sin 4x + \ldots$$

If a series of drawing is made (Fig. 2.9) showing, successively, the first term in this series, the sum of the first two terms, the sum of the first three terms, etc., we can gain some insight into the fascinating versatility of sines and cosines, which are able to cooperate in reproducing almost any conceivable function. It is especially interesting that discontinuous functions can be represented in this way.

---

*If a function f(x) is unchanged in value when x is replaced by $-x$, the function is said to have *even* symmetry about the origin. If the function merely changes its sign on such a replacement, it is said to have *odd* symmetry about the origin. The functions x, $\sin x$, x^3, and $\tan^{-1} x$ are odd functions, while x^2 and $\cos x$ are even. It is easy to see that the integral of the product of any even function with any odd function must vanish if the limits of integrationn lie of either side of the origin and at equal distances from the origin. This is a simple example of the use of symmetry operations in evaluating integrals. We shall see that in quantum mechanics extensive use is made of many kinds of operators for this prpose.

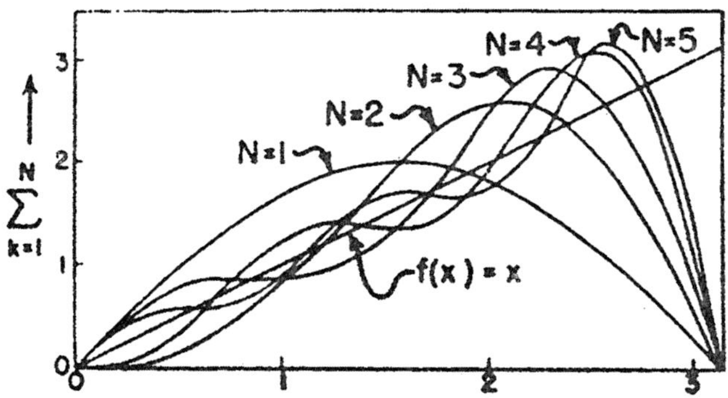

Fig. 2.9. Comparison of the incomplete Fourier series, $-\sum_{(k=1)}^{N} (-1)^k (2/k) \sin kx$ with the function $f(x) = x$. Note the improvement in the approximation as N increases from 1 to 5. (Curves are plotted only for x between 0 and π; similar curves are obtained if x lies between $-\pi$ and 0, but the ordinates negative.)

Other kinds of Fourier series are also possible. In the interval, $0 < x < \pi$, arbitrary functions can be expanded in terms of either sines or cosines alone

$$f(x) = \sum_{n=0}^{\infty} a_n \cos nx \tag{2.24}$$

$$f(x) = \sum_{n=0}^{\infty} b_n \sin nx \tag{2.25}$$

The student should show that for these two kinds of series,

$$a_n = (2/\pi) \int_0^{\pi} f(x) \cos nx \, dx; \; a_0 = (1/\pi) \int_0^{\pi} f(x) \, dx \tag{2.26}$$

$$b_n = (2/\pi) \int_0^{\pi} f(x) \sin nx \, dx \tag{2.27}$$

The interval over which the series is valid can also easily be changed merely by introducing a scaling factor. Thus the expansion

$$f(x) = \sum_{n=0}^{\infty} a_n \cos (n\pi x/l) + b_n \sin (n\pi x/l) \tag{2.28}$$

is valid in the range $-l < x < l$.

Other Series

A great many other sets of mutually orthogonal functions can serve as basis for the expansion of arbitrary functions over various intervals.

Problems

Find the Fourier expansions of some of the following functions and show by means of a graph how in each instance the sum of the first few terms begins to approach the true function.

(1) $f(x) = \sin^2 x$,

(i) in the range $-\pi$ to π

(ii) as a sine series (cf. Equation (2.25)) in the range 0 to π

(2) $f(x) = x$ as a cosine series (cf. Equation (2.24)) in the range 0 to π

(3) $f(x) = \begin{cases} x + 1 \text{ from } -\pi \text{ to } 0 \\ x - 1 \text{ from } 0 \text{ to } \pi \end{cases}$

(4) $f(x) = \begin{cases} 1 \text{ from } -\pi \text{ to } 0 \\ x - 1 \text{ from } 0 \text{ to } \pi \end{cases}$

(5) $f(x)=x^2$

(i) from $-\pi$ to π

(ii) as a sine series in the range 0 to π

The Fourier Integral

Because of the relationship between $e^{\pm i\theta}$ and $\cos\theta$ and $\sin\theta$ (Section *B* of this chapter), the Fourier series can be written in the equivalent form

$$f(x) = \sum_{n=-\infty}^{\infty} A_n e^{inx} \tag{2.29}$$

where

$$A_n = \frac{1}{2\pi} \int_{-\pi}^{\pi} f(x)\, e^{-inx}\, dx \tag{2.30}$$

Similarly Equation (2.28) can be written

$$f(x) = \sum_{n=-\infty}^{\infty} A_n e^{in\pi x/l} \tag{2.31}$$

where

$$A_n = \frac{1}{2l} \int_{-l}^{l} f(x) e^{-in\pi x/l}\, dx \tag{2.32}$$

It is interesting to see what happens to this representation when we let l approach infinity, so that the function $f(x)$ is represented over the entire range of the variable x, from $-\infty$ to ∞. Let us make the substitution $k = n\pi/l$. Note that if l is large, the new quantity k changes in small increments, $\Delta k = \pi/l$, when n is changed by the increment $\Delta n = 1$. Thus as $l \to \infty$, k becomes a continuous variable and we can write the coefficients A_n as ordinary functions of k $A(k)$. It is also convenient to write

$$lA(k)/\pi = g(k)/\sqrt{2\pi}$$

If, therefore, we set $1 = \Delta n$ in Equation (2.31), we may write

$$f(x) = \lim_{l \to \infty} \sum_{k=-\infty}^{\infty} A_n e^{in\pi x/l}\, \Delta n$$

$$= \lim_{\Delta k \to 0} \sum_{k=-\infty}^{\infty} \frac{lA(k)}{\pi}\, e^{ikx}\, \Delta k$$

or

$$f(x) = \frac{1}{\sqrt{2\pi}} \int_{-\infty}^{\infty} g(k)\, e^{ikx}\, dx \tag{2.33}$$

where

$$g(k) = \frac{1}{\sqrt{2\pi}} \int_{-\infty}^{\infty} f(x)\, e^{-ikx}\, dx \tag{2.34}$$

Equation (2.33) is known as a *Fourier integral* and $f(x)$ and $g(k)$ are said to be *Fourier transforms* of one another.

Problems

(1) Prove by direct substitution of (2.33) into (2.34) that $g(k)$, the Fourier transform of $f(x)$ in Equation (2.35), is very large when $k = 2\pi/\lambda_1$, $2\pi/\lambda_2$, $2\pi/\lambda_3$,..., and that the corresponding values of $g(k)$ are in the ratio $A_1 : A_2 : A_3$:...

(2) Find the Fourier transform of the function

$$f(x) = \begin{cases} 0; x < -\pi \\ x; -\pi < x < \pi \\ 0; x > \pi \end{cases} \tag{2.35}$$

[Answer : $g(k) = 2(k\pi \cos k\pi - \sin k\pi)/(i\sqrt{2\pi}\, k^2)$]

(3) Find the Fourier transform of the function

$$f(x) = \begin{cases} 0; x < -n\lambda \\ \sin 2\pi x/\lambda; -n\lambda < x < n\lambda \\ 0; x > n\lambda \end{cases}$$

[Answer : $g(k) = 2i\sqrt{2\pi}\,\lambda(k^2\lambda^2 - 4\pi^2)^{-1} \sin nk\lambda$]

In problem (3), $f(x)$ represents a train of monochromatic waves of finite duration. The Fourier transform in this instance shows that such a wave does not appear to be monochromatic when viewed through a spectroscope. It is seen, however, that the spectral intensity (as given by $|g(k)|^2$) is large only when $k\lambda \approx 2\pi$. If we write $k = 2\pi(1 + e)/\lambda$, where e is a new variable, we find that

$$|g(k)|^2 \propto \frac{\sin^2 2\pi n e}{(2e + e^2)^2}$$

This function has a series of peaks at $e = 0, \pm 3/n, \pm 5/n$,..., whose heights are respectively in the ratio $\pi^2/16{:}1/36{:}1/100$:... (see Fig. 2.10). Therefore most of the spectral intensity lies in the first peak, corresponding to k values lying within about $\pm\ (1/n)\ (2\pi/\lambda)$ of the mean value of k at $2\pi/\lambda$. The wave lengths in the finite "monochromatic" wave train, as measured in a spectroscope, are thus mostly distributed over a range between

$$\frac{\lambda}{1 + \frac{1}{n}} \text{ and } \frac{\lambda}{1 - \frac{1}{n}}$$

If $n = 100$, the light (as analyzed by the spectroscope) will appear to be made up of a distribution of wave lengths differing by as much as $\pm$ 1% from the mean value. It is clear that no light beam of finite duration can, on spectroscopic analysis, act as if it were truly monochromatic.

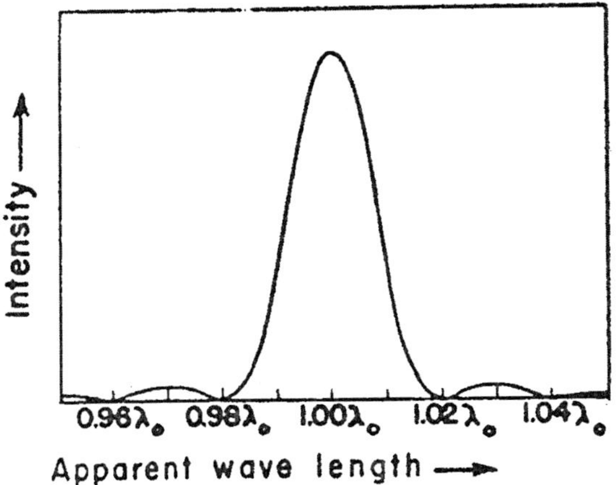

Fig. 2.10. Square of the Fourier transform of a wave train containing 200 oscillations whose wave length is λ_0. If this finite wave train is passed through a spectroscope, the intensities observed at different settings of the wave length scale of the spectroscope will be proportional to the ordinate.

ORDINARY DIFFERENTIAL EQUATIONS

Since some of the laws of quantum mechanics are expressed in the form of differential equations, it is important that we have a clear idea of just what a differential equation represents. A differential equation is simply an equation in which the derivatives of a function appear. If the function has only one independent variable, the differential equation is called an *ordinary differential equation.* For instance

$$dy/dx = x - y \qquad (2.36)$$

is an ordinary differential equation in which $y(x)$ is a function that depends on the independent variables, x. A differential equation that contains at most first derivatives is called a *first order differential equation*, while if the highest derivative it contains is a second derivative, it is called a *second order differential equation.* Most of the differential equations that we shall encounter in quantum chemistry are second order differential equations.

A differential equation is said to have been solved when all of the functions that satisfy it have been found. Every differential equation may be satisfied by a vast family of functions and it is often not easy to find out what these functions are. Furthermore, in many practical applications we seek a particular member of this family that satisfies certain specified conditions. Therefore the solution of differential equations can require a high order of mathematical skill and perseverance. Nevertheless, it is important to realize that there is nothing inherently obscure about what a differential equation stands for. One can, in fact, often gain some idea of the general appearance of the solutions of a differential equation by relatively simple methods.

Ordinary First Order Differential Equations

We shall now describe a simple method by which approximate graphical solutions of first order ordinary differential equations can usually be drawn. These equations can almost always be thrown into the form

$$dy/dx = f(x, y) \tag{2.37}$$

where $f(x, y)$ is a known function. According to this equation, for every value of x and y there is definite value of the slope, dy/dx. If, therefore, we set up x and y axes in a plane, a slope can be assigned to each of the points in this plane; that is, a short, straight line segment having the slope $f(x, y)$ and passing through the point (x, y) is a piece of one solution of the differential equation. If a sufficient number of such short line segments are drawn on the (x, y) plane, it should be possible to join them and so obtain all of the smooth curves that are the solutions of the differential equation.

Let us apply this method to the solution of Equation (2.36). We shall set

$$x - y = a \tag{2.38}$$

where $a = 0, \pm\frac{1}{2}, \pm 1, \pm\frac{3}{2}, \ldots$ Equation (2.38) represents a series of parallel straight lines in the (x, y) plane, and along each of these lines we may draw a set of short line segments all having the same slope, a. The result is shown in Fig. 2.11A. It is easy to see that these segments can be joined to give a family of smooth curves as shown in Fig. 2.11B.

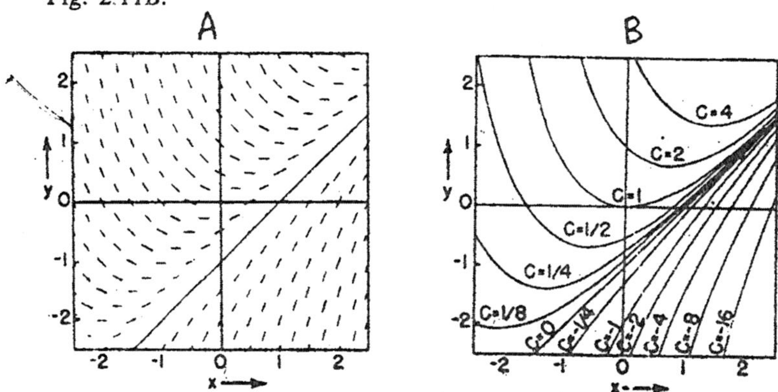

Fig. 2.11. Solutions of the differential equation $dy/dx = x - y$.
(a) Segments of the solutions passing through various points in the x-y plane as derived directly from the differential equation. Note that the slopes are identical along the lines $x = y + a$.
(b) Exact solutions, $y = x - 1 + Ce^{-x}$, corresponding to various values of the constant C. The general shapes of these solutions could be deduced directly from the segments drawn in A.

Problems

(1) The exact solution of Equation (2.36) is found to be $y = x - 1 + Ce^{-x}$, where C is an arbitrary constant. Verify that these solutions have the shapes shown in Fig. 2.11B, different values of the constant C giving different members of the family.

(2) Using the graphical method just described, find the general appearance of the family of solutions of the following differential equations :

(a) $dy/dx = x - \sin y$

(b) $dy/dx = xy + e^{xy}$

(c) $(dy/dx)^2 - (x + y \sin y)dy/dx + xy \sin y = 0$

Ordinary Second Order Differential Equations

A similar, though somewhat more qualitative, graphical method can be used with ordinary second order differential equations. Equations of this type can usually be thrown into the form

$$d^2/dx^2 = f(x, y, dy/dx) \tag{2.39}$$

The second derivative of a function is merely the rate at which the slope of the function changes when the independent variable is varied. If the slope of a function is not constant, the graph of the function must be curved; the value of d^2y/dx^2 is thus a measure of the "curvature" of the solution. Therefore a second order differential equation can be regarded as an expression for the *curvature* of its solutions, just as a first order differential equation may be regarded as an expression for the *slope* of its solutions.

Second order differential equations give us no information concerning the slope of the solution passing through a given point (x, y). Therefore this slope can be chosen arbitrarily. Accordingly, an infinite number of solutions, all differing in slope, can pass through each point in the x-y plane. Evidently a second order differential equation has an even greater multiplicity of solutions than we found for first order differential equations, for which only one solution passing through each point in the x-y plane. This greater multiplicity expresses itself through the occurrence of *two* arbitrary constants in the solutions of all second order differential equations.

Quantum mechanics is concerned only with second order differential equations having the special form,

$$P(x)d^2y/dx^2 + Q(x)dy/dx + R(x)y = 0 \tag{2.40}$$

where $P(x)$, $Q(x)$ and $R(x)$ are specified functions of x. Such an equation is called a *linear* second order differential equation, because the operator $P(x)d^2/dx^2 + Q(x)d/dx + R(x)$ is a linear operator. Linear differential equations have the useful property that, if $y = f(x)$ is a solution, then $y = Af(x)$ must also be a solution, where A is any constant. (The student can verify this by direct substitution in (2.40). Therefore, one of the two arbitrary constants appearing in the general solution of a linear second order differential equation merely eultiplies the function as a whole.

Now let us apply the graphical method to the linear second order differential equation

$$d^2y/dx^2 = -y \tag{2.41}$$

According to this equation, when y is positive the solution has a slope that decreases as x increases. On the other hand, if y is negative, the slope increases with increasing x. In short, all of the solutions are curved downward in the upper half of the x-y plane, and they are curved upward in the lower half of the x-y plane. Furthermore, the curvature of the solution passing through a given point is greater, the greater the value of the ordinate of the point. And when the solutions pass through the x-axis ($y = 0$) they are not curved at all; that is, points lying on the x-axis are always points of inflection.

It is easy to see that the solutions of (2.41) must oscillate from one side of the x-axis to the other as x is varied, because as long as we are on one side of the x- axis, the curvature is such that the solution tries to get to the other side, and once it gets to the other side, the solution will not be happy until it returns to the original side. Thus, by simply looking at the differential equation, it is possible to deduce the general behavior of the solutions.

In Fig. 2.12 a series of solutions of equation (2.41) is drawn, all of which pass through the point (0, 1), but with different slopes. It is readily verified by substitution in (2.41) that the family of solutions passing through this point has the general form $y = A \sin (x + b)$

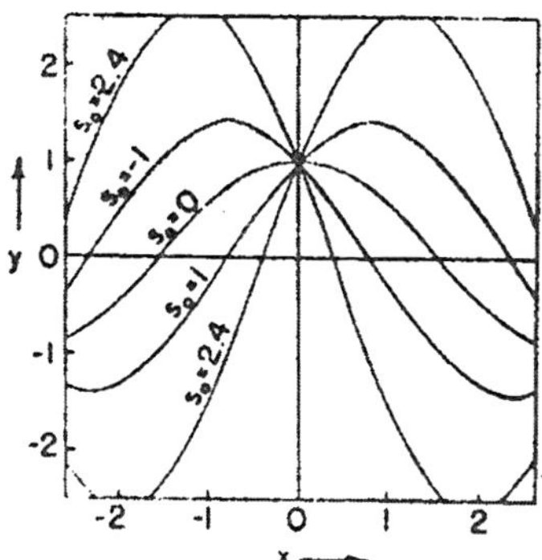

Fig. 2.12. Solutions of the differential equation $d^2y/dx^2 = -y$ passing through the point (0, 1) and having the indicated slopes s_0 at this point. Note tendency of the solutions to curve toward the x-axis.

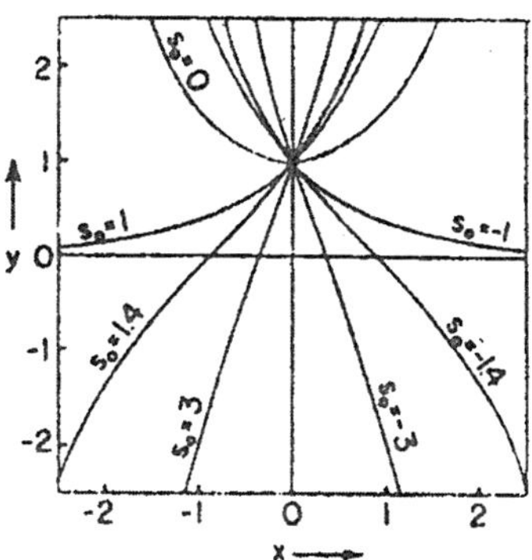

Fig 2.13. Solutions of the differential equation $d^2y/dx^2 = y$ passing through the point (0, 1) and having the indicated slopes s_0 at this point. Note tendency of the solutions to curve away from the x-axis.

where A and b are constants which are related by the condition $A \sin b = 1$. These solutions are, of course, exactly of the form inferred by means of the arguments in the previous paragraph.

Next let us consider the equation

$$d^2y/dx^2 = y \tag{2.42}$$

Using arguments similar to those just given, we can see that the solutions of this equation do not oscillate at all. In fact, they tend to "explode" since a large positive value of y tends to make the slope increase as x increases, so that if dy/dx is positive, y will tend to assume even larger values with increasing x, whereas if dy/dx is negative, y will increase without limit as x decreases. Similar explosive behavior is found for negative values of y, regardless of the sign of dy/dx. If we restrict ourselves to solutions which pass through the point (0, 1) we can easily see that nearly every solution explodes with both increasing and decreasing values of x (see Fig. 2.13); in some cases the explosion is toward positive values of y; in others, it is toward negative values. In only two instances do the solutions fail to explode in *both* the positive and negative x-directions. In these solutions y (and hence the curvature) approaches zero in such a way that the solution becomes asymptotic to either the positive or the negative x-axis. It is readily verified that the exact solutions of (2.42) which pass through the point (0, 1) are $y = Ae^{-x} + (1 - A)e^{x}$. The two asymptotic solutions result when $A = 0$ and 1. All other values of A result in "explosions" in both positive and negative x- directions.

The oscillatory and explosive behaviors described above are prominent features of the solutions of the differential equations of quantum mechanics. We shall find in Chapter 6 that energy quantization results from the fact that explosive solutions are not acceptable and that the asymptotic behavior is required at great distances from most systems of chemical interest.

Problems

(1) Describe the general appearance of the solutions of the differential equations

$$d^2y/dx^2 + (A - x^2)\,y = 0$$

which pass through the point (0, 0) and have slope unity at that point. (Do this by assuming initially that $A = 0$, and then show what happens to the solutions as A is increased. It will be found that for certain values of A the solutions will approach the x-axis asymptotically as $x \to \infty$ and as $x \to -\infty$. for all other values of A the solutions are of the explosive variety. This equation occurs in the quantum mechanics treatment of the harmonic oscillator and the asymptotic behaviour occurs only when the energy is quantized according to the well-known rule, $E = h\nu(n + 1/2)$.

(2) Legendre's equation of order one is

$$d^2y/dx^2 = [2x(dy/dx) - 2y]/(1 - x^2)$$

(a) Show that the only solutions of the equation that pass through the origin are those having the form $y = Ax$, where A is an arbitrary constant.

(b) Show that the solutions that passes through the point (0, 1) and has $dy/dx = 0$ becomes negatively infinite $x = \pm 1$.

(c) Show that all other solutions passing through the point (0, 1) also become negatively infinite at $x = \pm 1$.

(d) Show that all solutions passing $(0, a)$ become negatively infinite at $x = \pm 1$ if $a > 0$ and positively infinite at $x = \pm 1$ if $a < 0$.

PARTIAL DIFFERENTIAL EQUATIONS

When there is more than one independent variable in a system, differential equations containing partial derivatives will generally be obtained. In quantum mechanics we must frequently deal with linear second order partial differential equations; if there are two independent variables, these equations have the general form

$$P(x, y)\frac{\partial^2 f}{\partial x^2} + Q(x, y)\frac{\partial^2 f}{\partial x \partial y} + R(x, y)\frac{\partial^2 f}{\partial y^2}$$

$$+ S(x, y)\frac{\partial f}{\partial x} + T(x, y)\frac{\partial f}{\partial y} + U(x, y) f = 0 \qquad (2.43)$$

where P, Q, R, S, T, and U are given functions of the independent variables. If there are more than two independent variables the differential equations are of similar form except that there are additional terms involving first and second partial derivatives of the other variables.

Geometrical Interpretation of Partial Differential Equations

It is possible to discuss the geometrical significance of partial differential equations in a manner similar to that outlined above for ordinary differential equations. Evidently the curvature of the solutions of (2.43) with respect to any one of the variables is determined by the other curvatures and slopes appearing in the equation. An enormous variety of solutions is possible because even if there are only two variables, x and y, acceptable solutions will be obtained on giving arbitrary values to the function, to two of its slopes, $\partial f/\partial x$ and $\partial f/\partial y$, and to two of its curvatures, say $\partial^2 f/\partial y^2$ and $\partial^2 f/\partial x \partial y$.

For instance, consider the relatively simple second order equation

$$\frac{\partial^2 u}{\partial x^2} + \frac{\partial^2 u}{\partial y^2} = 0 \qquad (2.44)$$

Let us make a three dimensional plot of the solutions, $u(x, y)$, against the values of x and y. This plot will give a surface. Consider the intersection of this surface with the plane $x = 0$. The intersection of the plane with the surface result in a curve, $u(0, y)$, which we shall denote by $f(y)$. According to Equation (2.44), the solutions have merely to satisfy the condition

$$\partial^2 u/\partial x^2 = -\partial^2 u/\partial y^2 \qquad (2.45)$$

in the immediate vicinity of the y-u plane. The function $f(y)$ can have *any shape whatsoever,* as long as it has a second derivative, and furthermore the values of $\partial u/\partial x$, which give the slope of the surface in a direction normal to the y-u plane, are *completely arbitrary*. Thus an enormous number of different surfaces are consistent with Equation (2.44). In fact, as we shall see in studying the shapes of vibrating membranes, all of the shapes that can be assumed by a stationary stretched elastic membrane are consistent with Equation (2.44).

The student might gain the impression from the above discussion that, because of the enormous diversity in the shapes of their solutions, partial differential equations do not tell us very much about a system. This, however, is not the case. In practical applications of partial differential equations, we almost always known something about the solutions before we begin. For instance, in vibrating bodies we know that certain portions are held rigidly in place, or that certain portions have no stresses acting on them. In quantum mechanics we know that the solutions are well behaved. These conditions—which are called *boundary conditions*—are usually sufficient to limit the solutions very drastically, so that the solutions provide us with detailed description of the possible behaviour of the systems.

Problems

(1) Show that the following functions are solutions of Equation (2.44)

(a) $u = \log(x^2 + y^2)$

(b) $u = x/(x^2 + y^2)$

(c) $u = Af(x + iy) + Bg(x - iy)$

where A and B are arbitrary constants, $i = \sqrt{-1}$, and f and g are any functions that have second derivatives. Show that the solutions (a) and (b) can be written in the form (c).

(2) Show that the differential equation

$$\frac{\partial u}{\partial x} + \frac{\partial u}{\partial y} = 0$$

has the general solution $u(x, y) = Af(x - y)$, where A is an arbitrary constant and f is any function in which x and y always appear in the combination $(x - y)$ (e.g., $u = (x - y)^3$; $u = \sin(x - y)$; $u = \exp(x - y)$). Draw a sketch of the solution that, for $y = 0$, has the shape $u(x, 0) = \exp(-x^2)$.

The Solution of Partial Differential Equations by the Method of Separation of Variables

Consider the differential equation

$$\frac{\partial u}{\partial x} - \frac{\partial u}{\partial y} = 0 \qquad (2.46)$$

Let us see what happens if we assume that the solution has the form

$$u(x, y) = X(x)\, Y(y) \qquad (2.47)$$

where $X(x)$ is a function involving only x and $Y(y)$ involves only y. On differentiating we find

$$\partial u/\partial x = Y\, dX/dx;\ \partial u/\partial y = X\, dY/dy \tag{2.48}$$

Total derivatives are written on the right hand sides of these two equations because each of the differential functions contains only one variable. On substituting (2.48) into (2.46) we find

$$Y\, dX/dx - X\, dY/dy = 0 \tag{2.49}$$

Dividing through by $u = XY$, we obtain an equation in which the variables have been "separated" into different terms

$$\frac{1}{X(x)}\frac{dX(x)}{dx} - \frac{1}{Y(y)}\frac{dY(y)}{dy} = 0 \tag{2.50}$$

It is easy to see that each of the two terms appearing on the left hand side of (2.50) is constant independent of both x and y. The first term, $(1/X)dX/dx$, clearly depends only on x and is independent of y, whereas the second term, $(1/Y)dY/dy$, does not depend on x. Therefore, if we vary x in Equation (2.50), the second term on the left must remain constant. Since the equation as a whole must be true for all values of x, this means that the first term on the left is also independent of x. Similarly, as y is varied, the first term on the left cannot vary, so the second term must also be independent of y. In other words, both $(1/X)dX/dx$ and $(1/Y)dY/dy$ are constant. Furthermore, the two terms must be equal to one another in order that their difference vanish. Thus we arrive at the two ordinary differential equations

$$(1/X)dX/dx = c \tag{2.51}$$

$$(1/Y)dY/dy = c \tag{2.52}$$

where c is an arbitrary constant. These equations have the solutions

$$X = Ae^{cx} \tag{2.53}$$

$$Y = Be^{cy} \tag{2.54}$$

where A and B are arbitrary constants. Combining A and B into a single arbitrary constant, $D = AB$, we obtain for the solution of (2.46)

$$u(x, y) = De^{c(x+y)} \tag{2.55}$$

The Method outlined above does not always work. In some instances the solution obtained does not fit the boundary conditions that have been set for the problem. More often the coefficients P, Q, R, S, T, or U of Equation (2.43) contain the variable in such combinations that they cannot be separated from each other. In either case it is sometimes possible to transform the differential equation by using a new coordinate system and so obtain an equation which can be separated. Unfortunately, as Robertson and Eisenhart have shown, separation of variables is possible for only a few general types of equations occurring in quantum mechanics. Almost all of the more complex chemically interesting quantum mechanical problems give equations that do not belong to one of these types, so that the method of separation of variables is not applicable. It is this fact more than any other that stands in the way of progress in the rigorous application of quantum mechanics to chemistry and in the precise *a priori* calculation of the properties of chemical systems. Nevertheless we shall find frequent occasion to use this method in the solution of simpler problems.

Problem

Can the method of separation of variables be used with the equation

$$\frac{\partial^2 u}{\partial x^2} + \frac{\partial^2 u}{\partial y^2} + \frac{u}{\sqrt{x^2 + y^2}} + u = 0$$

Show that the transformation $x = r \cos \theta$, $y = r \sin \theta$ will give a differential equation that can be separated, and find the resulting total differential equations.

DETERMINANTS

A *determinant* is an arrangement of quantities or *elements* A_{ij} in rows and columns in which the number of rows is equal to the number of columns

$$\begin{vmatrix} A_{11} & A_{12} & A_{13} & \ldots & A_{1n} \\ A_{21} & A_{22} & A_{23} & \ldots & A_{2n} \\ A_{31} & A_{32} & A_{33} & \ldots & A_{3n} \\ \cdot & \cdot & \cdot & & \cdot \\ \cdot & \cdot & \cdot & & \cdot \\ \cdot & \cdot & \cdot & & \cdot \\ A_{n1} & A_{n2} & A_{n3} & \ldots & A_{nn} \end{vmatrix} \tag{2.56}$$

The number of rows or columns, n, is called the *order* of the determinant.

Each element in a determinant has a *minor,* which is defined as the determinant remaining when the row and column containing the given elements are removed from the original determinant. The minors of the elements in an nth order determinant are therefore determinants of order $(n - 1)$.

The *value* of a determinant is defined as the quantity obtained in the following way. There an n ! different ways of choosing one and only one element from each row and each column in any n^{th} order determinant. Form the n ! different products each containing n elements chosen in this way. Arrange the elements A_{ij} in each of these products so that their first subscripts are in their natural order, 1, 2, 3, ..., n. In all but one of the products the second subscripts will not occur in their natural order; find out how many permutations of the elements in each product are required in order to bring the second subscripts into their natural order, 1, 2, 3, ..., n. If this number of permutations is even, give the product a positive sign, and if it is odd, give it a negative sign. Then the value of the determinant is simply the algebraic sum of all of the products.

In order to illustrate this procedure, consider the third order determinant

$$\begin{vmatrix} A_{11} & A_{12} & A_{13} \\ A_{21} & A_{22} & A_{23} \\ A_{31} & A_{33} & A_{33} \end{vmatrix} \tag{2.57}$$

The 3! = 6 different products obtained from this determinant are, when written so that their first subscripts fall in the order, 1, 2, 3

$$A_{11} A_{22} A_{33} \ (p = 0) \qquad A_{11} A_{23} A_{32} \ (p = 1)$$

$$A_{11} A_{21} A_{33} \ (p = 1) \qquad A_{12} A_{23} A_{31} \ (p = 2)$$

$$A_{13} A_{22} A_{31} \ (p = 1) \qquad A_{13} A_{21} A_{32} \ (p = 2)$$

The number of permutations, p, required to bring the second subscripts into the order 1, 2, 3 in each case is indicated beside each product. The value of the determinant is thus

$$A_{11} A_{22} A_{33} - A_{12} A_{21} A_{33} - A_{13} A_{22} A_{31} - A_{11} A_{23} A_{32} + A_{12} A_{23} A_{31} + A_{13} A_{21} A_{32} \tag{2.58}$$

In general we may write for the value of a determinant

$$\sum (-1)^p A_{1k_1} A_{2k_2} A_{3k_3} \cdots A_{nk_n} \tag{2.59}$$

where p is the number of permutations required to bring $k_1, k_2, k_3, \ldots, k_n$ into the order 1, 2, 3, ..., n and the sum is over all permutations of the k_i's.

It is easy to show that the value of a determinant can also be found in the following way : Multiply each of the elements in the first row or column by its minor. Give the product a plus sign if the element is first, third, fifth, ... in the row or column, and a minus sign if it is second, fourth, sixth, ... Add all of the products. The result is the value of the determinant. Thus we have for the determinant (2.57).

$$A_{11} \begin{vmatrix} A_{22} & A_{23} \\ A_{32} & A_{33} \end{vmatrix} - A_{21} \begin{vmatrix} A_{12} & A_{13} \\ A_{32} & A_{33} \end{vmatrix} + A_{31} \begin{vmatrix} A_{12} & A_{13} \\ A_{22} & A_{23} \end{vmatrix} \tag{2.60}$$

which gives the same result as (2.58).

The evaluation and manipulation of determinants is greatly simplified by taking note of the following properties of determinants (all of which are easily proven by means of the definitions given above).

(1) The value of a determinant merely changes sign when two rows or two columns are interchanged.

(2) If every element in a row or column is zero, the value of the determinant is zero.

(3) If two rows are equal, or if two columns are equal, the determinant is zero.

(4) If all of the elements of the ith row are written as the sums of two numbers, $A_{ij} = B_{ij} + C_{ij}$, then the determinant can be written in the form

$$\begin{vmatrix} A_{11} & A_{12} & \cdots & A_{1n} \\ \vdots & \vdots & & \vdots \\ A_{i1} & A_{i2} & \cdots & A_{in} \\ \vdots & \vdots & & \vdots \\ A_{n1} & A_{n2} & \cdots & A_{nn} \end{vmatrix} = \begin{vmatrix} A_{11} & A_{12} & \cdots & A_{1n} \\ \vdots & \vdots & & \vdots \\ B_{i1} & B_{i2} & \cdots & B_{in} \\ \vdots & \vdots & & \vdots \\ A_{n1} & A_{n2} & \cdots & A_{nn} \end{vmatrix} + \begin{vmatrix} A_{11} & A_{12} & \cdots & A_{1n} \\ \vdots & \vdots & & \vdots \\ C_{i1} & C_{i2} & \cdots & C_{in} \\ \vdots & \vdots & & \vdots \\ A_{n1} & A_{n2} & \cdots & A_{nn} \end{vmatrix}$$

A similar relationship holds for the columns of a determinant.

(5) If one multiplies all terms in one row or one column by a constant, the determinant is multiplied by the same constant.

(6) If one multiplies all members of one row or one column by a constant adds the result to another row or column, the value of the determinant is unchanged.

A frequent used practical procedure for evaluating determinants of high order is to make all but one of the elements in one row equal to zero by means of property (6). The order of the determinant is in this way effectively reduced by one. The same reduction procedure can then be employed on the minor of the remaining nonzero element, and so on.

In the quantum chemical treatment of aromatic and conjugated hydrocarbons we shall encounter so-called *cyclic determinants.* An nth order cyclic determinant, C, is formed from n quantities $c_1, c_2, c_3, \ldots, c_n$ by rearranging their order in different rows in the following way

$$C = \begin{vmatrix} c_1 & c_2 & c_3 & \cdots & c_n \\ c_2 & c_3 & c_4 & \cdots & c_1 \\ c_3 & c_4 & c_5 & \cdots & c_2 \\ \cdot\cdot & \cdot\cdot & \cdot\cdot & \cdots & \cdot\cdot \\ c_n & c_1 & c_2 & \cdots & c_{n-1} \end{vmatrix}$$

It can be shown that the value of this determinant is given by

$$C = (-1)^{(n-1)(n-2)/2} f_1 f_2 f_3 \ldots f_n \tag{2.61}$$

where
$$f_k = \sum_{j=1}^{n} c_j x_k^{\,j-1}$$

$$kx_k = \exp(2k\pi i/n)$$

and
$$i = \sqrt{-1}$$

Problems

(1) Evaluate the determinant

$$\begin{vmatrix} 4 & 1 & 2 & 3 \\ 1 & 2 & 3 & 4 \\ 2 & 3 & 4 & 1 \\ 3 & 4 & 1 & 2 \end{vmatrix}$$

using the method of minors or Equations (2.61). [Answer : – 160.]

(2) Show that

$$\begin{vmatrix} 1 & 2 & 3 & \dots & n \\ 2 & 3 & 4 & \dots & 1 \\ 3 & 4 & 5 & \dots & 2 \\ \dots & \dots & \dots & \dots & \dots \\ n & 1 & 2 & \dots & (n-1) \end{vmatrix} = (-1)^{n(n-1)/2} \frac{n=1}{2} n^{n-1}$$

(3) Consider the equation

$$C_n \equiv \begin{vmatrix} x & 1 & 0 & 0 & \dots & 0 & 0 & 1 \\ 1 & x & 1 & 0 & \dots & 0 & 0 & 0 \\ 0 & 1 & x & 1 & \dots & 0 & 0 & 0 \\ 0 & 0 & 1 & x & \dots & 0 & 0 & 0 \\ \cdot & \cdot & \cdot & \cdot & \cdot & \cdot & \cdot & \cdot \\ 0 & 0 & 0 & 0 & \dots & 1 & x & 1 \\ 1 & 0 & 0 & 0 & \dots & 0 & 1 & x \end{vmatrix} = 0 \tag{2.62}$$

where C_n is an nth order cyclic determinant in which all but three elements in each row vanish. Show that the n roots of this equation are given by

$$x = -2\cos(2\pi k/n) \tag{2.63}$$

where $k = 1, 2, 3, \dots, n$.

(4) Let D_n be the nth order non-cyclic determinant defined by

$$D_n \equiv \begin{vmatrix} x & 1 & 0 & 0 & \dots & 0 & 0 & 0 \\ 1 & x & 1 & 0 & \dots & 0 & 0 & 0 \\ 0 & 1 & x & 1 & \dots & 0 & 0 & 0 \\ 0 & 0 & 1 & x & \dots & 0 & 0 & 0 \\ \cdot & \cdot & \cdot & \cdot & \dots & \cdot & \cdot & \cdot \\ 0 & 0 & 0 & 0 & \dots & x & x & 1 \\ 0 & 0 & 0 & 0 & \dots & 1 & 1 & x \end{vmatrix} \tag{2.64}$$

(Note that this determinant is identical with C_n in Problem (3) except in the first and last rows, where all but *two* of the elements vanish.)

(a) Show that

$$D_n = xD_{n-} - D_{n-2} \tag{2.65}$$

where

$$D_1 = x \tag{2.66}$$

and

$$D_2 = \begin{vmatrix} x & 1 \\ 1 & x \end{vmatrix} = x^2 - 1 \tag{2.67}$$

(b) Show that if one sets $x = 2\cos\theta = e^{i\theta} + e^{-i\theta}$, then

$$D_n = Ae^{in\theta} + Be^{-in\theta} \tag{2.68}$$

where A and B may depend on θ but not on n. [Hint : Use the method of induction, showing that if the equation holds for D_n it also holds for D_{n+1}, and that it must therefore hold for all values of n.]

(c) show that $A = e^{i\theta}/(e^{i\theta} - e^{-i\theta})$ and $B = -e^{-i\theta}/(e^{i\theta} - e^{-i\theta})$. [Hint : Compare (2.68) with (2.66) and (2.67).]

(d) Using these valued of A and B in ((2.68) show that

$$D_n = \frac{\sin(n+1)\,\theta}{\sin\theta} \tag{2.69}$$

(e) Show that the roots of the equation $D_n = 0$ occur at

$$x = 2\cos\,[k\pi/(n+1)] \tag{2.70}$$

where $k = 1, 2, 3, \ldots, n$.

3

The Classical Theory of Vibrations

Quantum mechanics can be adequately grasped only by understanding its mathematical structure. It is therefore helpful to learn some of the mathematics in the framework of a more familiar physical problem. The classical theory of the vibrations of elastic bodies furnishes an excellent training field. Not only is the mathematical treatment of vibrations closely similar to that of a large segment of quantum mechanics, but the theory of vibrations is of interest and value in its own right. Before going into the quantum mechanical theory we shall therefore find it useful to devote some time to the study of ordinary vibrations.

We are all familiar with the great variety of oscillations that occur when a hanging chain or a dish of water is disturbed. All material objects are in fact able to vibrate, though the frequencies are often too high and the amplitudes too low for us to notice the motion directly.

SIMPLE HARMONIC MOTION

Consider a point P which revolves counterclockwise on the circumference of a circle of radius A. Such an arrangement is shown in Fig. (3.1). Further assume that the point starts at zero time from position S and revolves with a *constant angular frequency* ω (radiant/second). We'll designate point Q as the horizontal projection of point P on the y-axis passing through the origin. As point P revolves, point Q oscillates up and down in what is called *simple harmonic motion. We may define simple harmonic motion as the projection of circular motion on the diameter of the circle.*

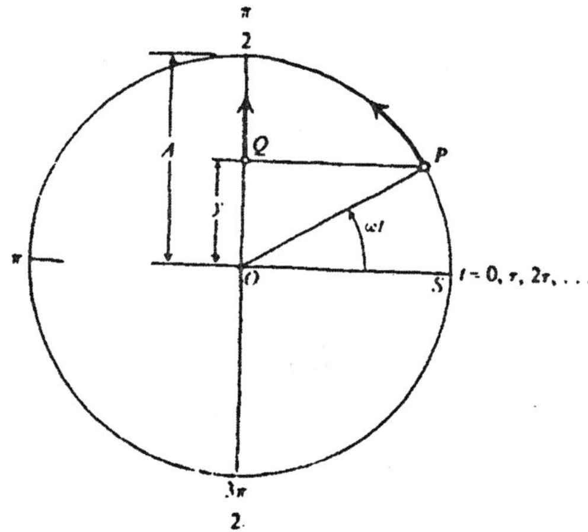

Fig. 3.1. Simple harmonic motion as a projection of circular motion on the diameter of the circle.

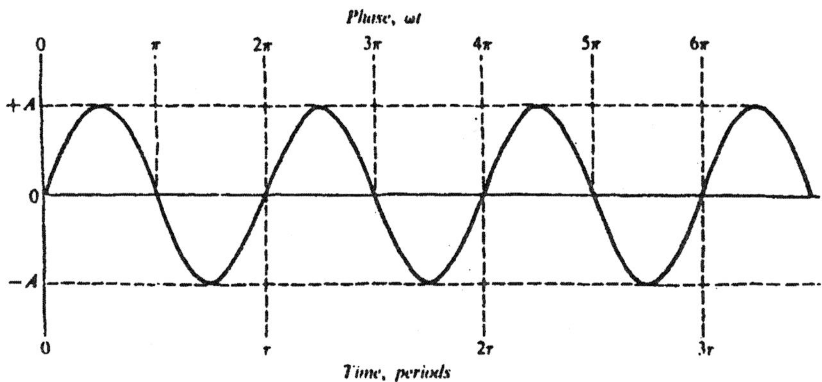

Fig. 3.2. Simple harmonic motion. Displacement y as a function of time t.

The motion of point Q repeats itself during each successive revolution. The period of time required for one revolution (or *cycle*) is called the period τ, and is give by

$$\tau = \frac{2\pi}{\omega}\frac{\text{rad/cycle}}{\text{rad/sec}} = \frac{\text{sec}}{\text{cycle}}. \tag{3.1}$$

It also follows that the *frequency* of revolution, ν, in cycles/second is given by

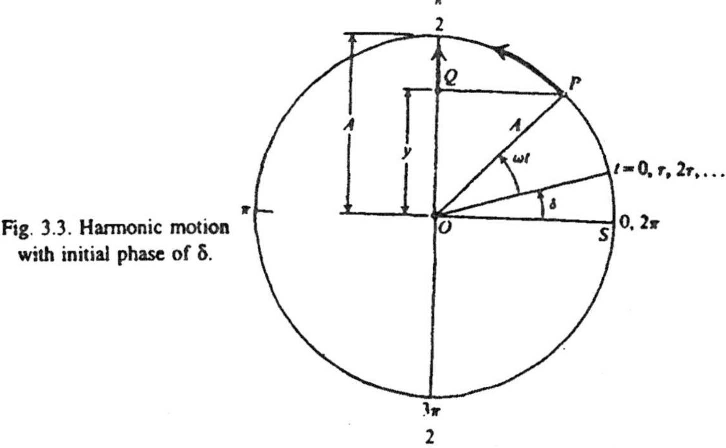

Fig. 3.3. Harmonic motion with initial phase of δ.

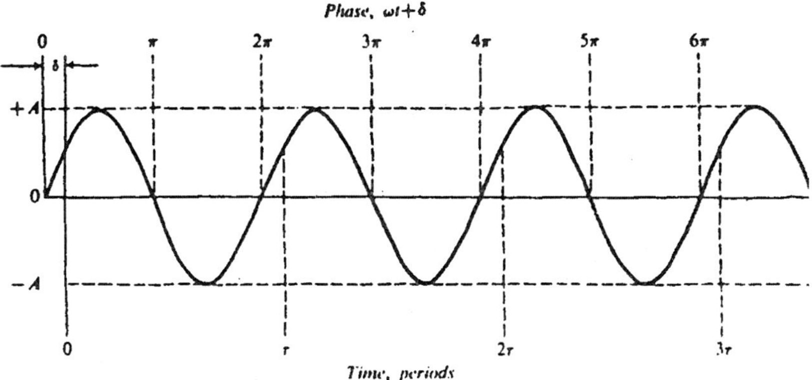

Fig. 3.4. Simple harmonic motion with phase displacement δ.

$$\nu = 1/\tau = \omega/2\pi,$$

or

$$\omega = 2\pi\nu. \tag{3.2}$$

In order to define mathematically the motion of point Q, which is undergoing simple harmonic oscillation, it is convenient to define the *displacement* y as the distance of point Q from the origin or center of oscillation. Thus, from Fig. (3.1), we have

$$y = A \sin (\omega t). \tag{3.3}$$

Equation (3.3) mathematically defines simple harmonic motion. A plot is shown in Fig. (3.2). The maximum displacement A is called the *amplitude,* and the quantity ωt is called the *phase.*

If, instead of starting the circular motion of point P as S, we start at some displaced angle δ, as shown in Fig. (3.3), the displacement is given by

$$y = A \sin (\omega t + \delta), \tag{3.4}$$

where the *phase* is now $\omega t + \delta$ and δ is referred to as the *initial phase* or *phase constant.* The displacement y is plotted as a function of t according to Eq. (3.4) in Fig. (3.4). At any point in time, the phases of the waves shown in Figs. (3.2) and (3.4) are different by the phase constant δ. the two waves are said to be *out of phase.*

The velocity of point Q at any time may be expressed as dy/dt. Differentiation of the more general form of the wave equation for harmonic motion, Eq. (3.4), gives

$$\frac{dy}{dt} = A\omega \cos (\omega t + \delta). \tag{3.5}$$

The acceleration of point Q, which is undergoing simple harmonic motion, is given by the relations

$$\frac{d^2y}{dt^2} = -A\omega^2 \sin (\omega t + \delta). \tag{3.6}$$

Combination of Eq. (3.4) and Eq. (3.6) to eliminate $A \sin (\omega t + \delta)$ gives

$$\frac{d^2y}{dt^2} = -\omega^2 y, \tag{3.7}$$

which is the *differential equation of harmonic motion.* Equation (3.7) is a more general description of harmonic motion than Eq. because it does not include dependence on the constants A and δ.

Energy of the Simple Harmonic Oscillators

Although kinetic energy T and potential energy V are periodically interconverted during the vibration of an isolated linear harmonic oscillator, their total remains constant and equal to E. That is,

$$E = T + V. \tag{3.8}$$

The potential energy V of the system is the amount of work done is displacing the mass from its equilibrium position to a displacement y, so that

$$V = \int_0^y -f\,dy = \int_0^y ky\,dy = ky^2/2. \tag{3.9}$$

Since $T = m\text{v}^2/2$, where v is the velocity of the mass, Eq. (3.8) may be written as

$$E = (m\text{v}^2/2) + (ky^2/2). \tag{3.10}$$

Although E is the same at any time during the oscillation, it is convenient to evaluate E at the point of maximum displacement, where $y = A$ and $\text{v} = 0$, whereupon Eq. (3.10) becomes

$$E = \tfrac{1}{2}\kappa A^2, \tag{3.11}$$

and the *total energy of the simple harmonic oscillator is shown to be directly proportional to the square of its amplitude.*

OSCILLATION OF A MASS ON A SPRING

In a physical sense, simple harmonic motion may also be defined as the *motion of a body which is constrained to move in a straight line, subject to a restoring force which is proportional to its displacement from the equilibrium position.* This can be proved as follows : Consider a mass which is oscillating in a vacuum and in the absence of gravity on a weightless spring as shown in Fig. (3.5). Further, assume that the spring is attached on the other end to a large immovable mass such as a wall. If y is the displacement from the equilibrium position, *Hooke's law* may be stated as

$$f = -ky, \tag{3.12}$$

where f is the restoring force and k is a characteristic constant of the spring called the *force constant.* The negative sign in Eq. (3.12) indicates that the force acts in opposition to the displacement; that is, it is a *restoring force.*

The velocity of the mass at any point is dy/dt, and the acceleration isd^{2y}/dt^2. According to Newton's second law of motion, the restoring force is also given by the relation :

$$f = ma = m\frac{d^2y}{dt^2}. \tag{3.13}$$

Eqs. (3.12) and (3.13) may be combines to eliminate f, thus we have

$$\frac{d^2y}{dt^2} = \frac{-k}{m}y, \tag{3.14}$$

which is of the same form as Eq. (3.7), the differential equation of linear harmonic motion, thus revealing that the mass on the spring is a linear harmonic oscillator. A comparison of the coefficients of y in Eqs. (3.7) and (3.14) relates the angular frequency ω to κ and m such that

$$\omega^2 = \kappa/m. \tag{3.15}$$

Furthermore, a satisfactory displacement equation for the mass must be given by Eq. (3.4), since Eq. (3.14) is derived through double differentiation of Eq. (3.4), so that

$$y = A \sin [(\kappa/m)^{1/2}t + \delta]. \tag{3.16}$$

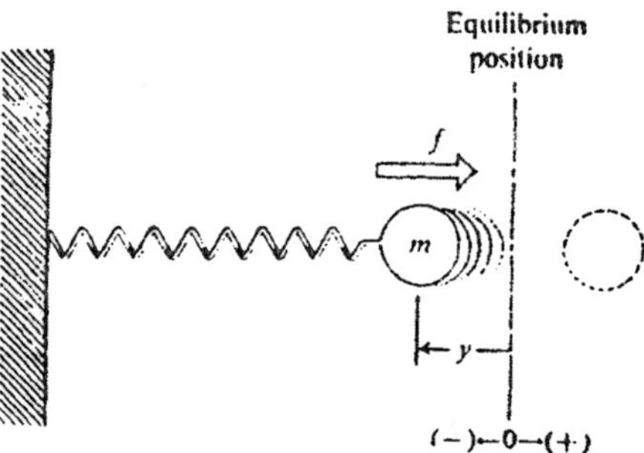

Fig. 3.5. Mass oscillating on a spring.

Let's now examine in details how we might obtain in a ***different*** fashion the displacement equation $y(t)$ by the direct solution of Eq. (3.14). Even though we have already shown above, through comparison of the differential Eqs. (3.7) and (3.14), that a satisfactory displacement equation is given by Eq. (3.4), we'll perform this additional exercise in order (a) to introduce an important technique which we'll use often in solving linear differential equations such as Eq. (3.14), and (b) to show that the sine form given by Eq. (3.4) is not the only possible form for the displacement equation.

In quantum mechanics, operator* are often (but not always) denoted by a *circumflex* above a letter. Typically operator symbols are $\hat{p}$, $\hat{H}$, and $\hat{T}$.

In using operators we'll often encounter operator equation of the type

$$\hat{p}\psi = p\psi,$$

which are called *eigenvalue equations,* where $\hat{p}$ is usually a differential operator, ψ is an eigenfunction of the operator. $\hat{p}$ and p is a number called the *eigenvalue.* An eigenvalue equation is solved by finding an eigenfunction ψ such that the operation of $\hat{p}$ on ψ yields the product of the original eigenfunction ties a constant. Usually more than one eigenfunction is found to satisfy a given operator, and more than one eigenvalue results.

Eq. (3.14) has the form of an eigenvalue equation where

$\dfrac{d^2}{dt^2}$ is the operator, $\dfrac{-k}{m}$ is the eigenvalue, and y is the eigenfunction.

In order to solve Eq. (3.14) we must find a function $y(t)$ such that when the second derivative of $y(t)$ is taken with respect to t, the result is the original function $y(t)$ multiplied by $(-\kappa/m)$.

A common method for solving eigenvalue equation is trial and error. For example, we know that the *second* derivative of a sine function produces a negative sine function. We might, then propose the general function

$$y = a \sin (bt + c) \tag{3.17}$$

as a possibly acceptable eigenfunction for the operator d^2/dt^2, where a, b, and c are constants. Let's perform the required operation in order to test the function;

$$\frac{dy}{dt} = ba \cos (bt + c), \quad \frac{d^2y}{dt^2} = -b^2 a \sin (bt + c),$$

or

$$\frac{d^2[a \sin (bt + c)]}{dt^2} = -b^2 [a \sin (bt + c)].$$

---

*The mathematical concept of operators. is discussed in chapter 2.

That is,

$$\frac{d^2y}{dt^2} = -b^2y \tag{3.18}$$

Comparison of Eq. (3.18) with Eq. (3.14) indicates that $y = a \sin(bt + c)$ is a satisfactory eigenfunction of the operator d^2/dt^2 if and only if

$$b^2 = \kappa/m.$$

Thus the displacement equation, Eq. (3.13), may be written as

$$y = a \sin[(\kappa/m)^{1/2} t + c] \tag{3.19}$$

which is identical to Eq. (3.16) a is the amplitude A, and c is the initial phase δ.

The frequency ν of vibration of the mass on the spring may be expressed by substituting $2\pi\nu$ for ω in Eq. (3.15), which yields :

$$\nu = \frac{1}{2\pi}\left(\frac{\kappa}{m}\right)^{1/2}. \tag{3.20}$$

The frequency given above is the only allowed *natural frequency*. Any other frequency forced upon the oscillating mass is said to be *anharmonic*.

Finally, it is important to note that functions other than the sine function may serve as satisfactory displacement equations for the linear harmonic oscillator. It can be shown easily, for example, that the functions

$$y = A \cos(\omega t + \delta) \tag{3.21}$$

and

$$y = e^{i(\omega t + \delta)}, \tag{3.22}$$

where $i^2 = -1$, are equally satisfactory eigenfunctions of the operator, d^2/dt^2 and are thus equally satisfactory displacement equations for the linear harmonic oscillator. In fact, we can shown that *any linear combination* of the sine, cosine, and complex-exponential displacement functions also satisfies the general differential equation of harmonic motion; and is, therefore, a valid description for simple harmonic motion.

TRAVELING WAVES IN A STRETCHED STRING

In order to more completely understand the wave properties of a moving particle, which in quantum mechanics may be represented by the properties of a traveling wave, it is necessary to examine in some detail the nature of the one-dimensional traveling wave which is produced in an infinitely long string under tension.

If one were to grasp one end of a long, taut and were to give the single vertical flip, a single pulse wave would be generated which would travel along the entire length of the Fig. (3.6). If, however, one were to continually oscillate the string end up and down at a fixed frequency, a traveling wave would be generated in the Fig. (3.7). Note that although the waveform travels *longitudinally* (x-direction) along the string, a given segment of string undergoes a periodic *transverse* (y-direction) displacement which is a function of both time and longitudinal position in the string.

In order to more completely analyze displacement as a function of time and longitudinal position, let's consider an infinitely long string which is under tension. Assume that

1. The string is uniform throughout its length.
2. The force of tension F acting axially in the string at any point is constant and is great enough so that it is not appreciably changed by the vibration.
3. The displacement y is small and is a smooth function of the longitudinal position x.

A section of such a taut string is shown in Fig. 3.8. The waveform is shown as moving from left to right at a *constant phase velocity* or *propagation velocity* in the x- direction, and the displacement y is a periodic function of both x and t.

Let's now analyze the forces acting on a differential length dL of the string at a specific moment in time, that is, *while holding time constant.* Let's look at the differential string segment dL shown at point a in Fig. 3.8. this section is magnified in Fig. 3.9. The force of tension F acts tangentialdy to the string at each end of the segment dL. The component force f_2 acting *down* on the segment at point 2 is given by the relations :

$$f_2 = F \sin \theta_2, \tag{3.23}$$

whereas the component force f_1 acting *up* on the segment at point 1 is given by the relation :

$$f_1 = F \sin \theta_1. \tag{3.24}$$

The net downward component of force which tends to *restore* the string segment to its equilibrium position is thus

$$f = f_2 - f_1 = F(\sin \theta_2 - \sin \theta_1). \tag{3.25}$$

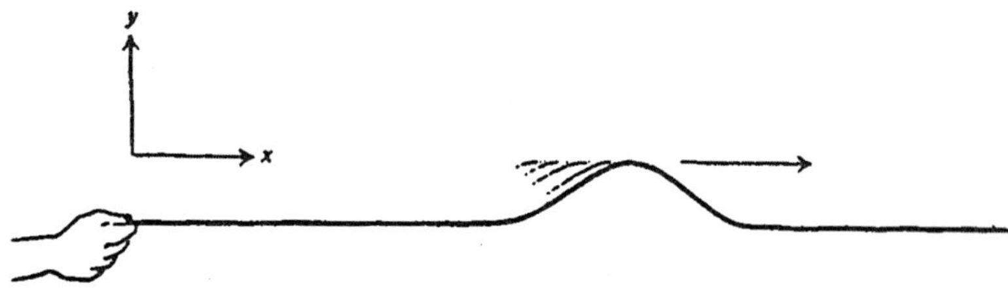

Fig. 3.6. A single pulse wave generated by a single impulse on the end of a taut string.

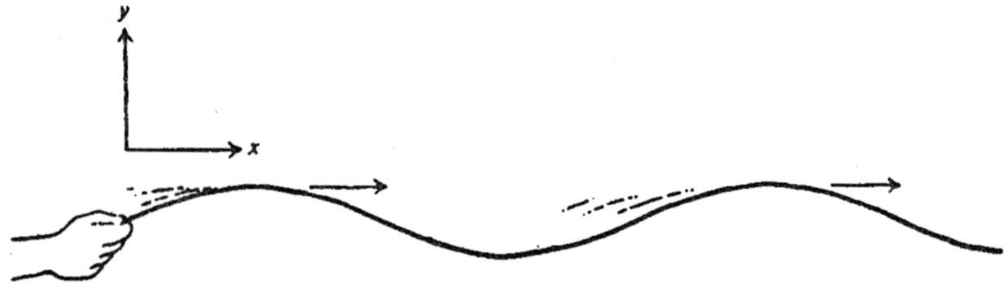

Fig. 3.7. A traveling wave generated by oscillating the end of a taut string.

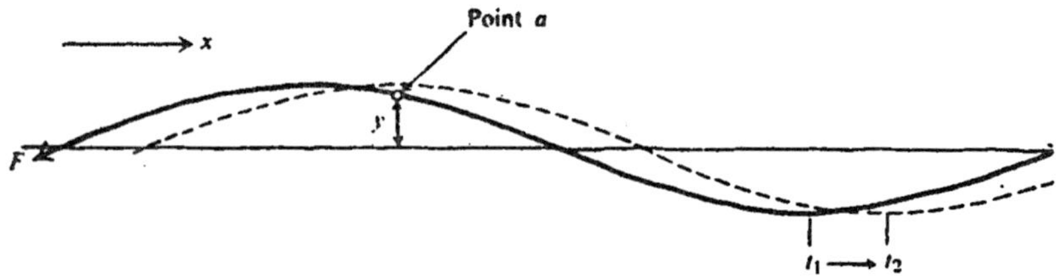

Fig. 3.8. Transverse displacement y as a function of time t and longitudinal position x in a traveling wave. The solid line shows the wave at time t_1, and the dashed line shows the wave at t_2, a short time later.

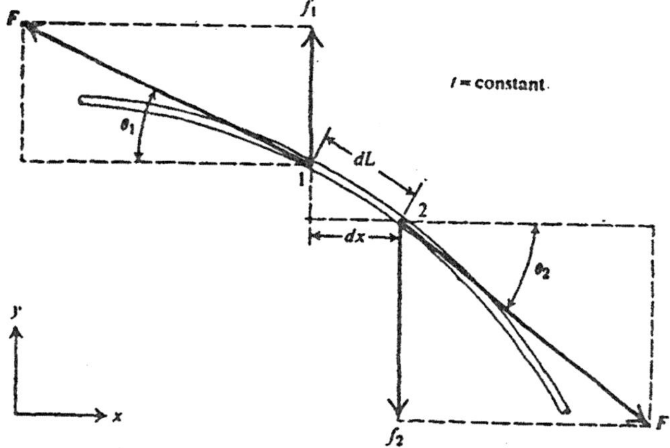

Fig. 3.9. The forces acting on a differential segment, *dL*, of a stretched string under vibration.

But if θ is small, which it is when the displacement is small, sin θ ≡ tan θ, so that

$$f = F(\tan\theta_2 - \tan\theta_1). \tag{3.26}$$

But tan θ is the slope of the wave at constant time, $(\partial y/\partial x)_t$, thus we have

$$f = F\left[\left(\frac{\partial y}{\partial x}\right)_2 - \left(\frac{\partial y}{\partial x}\right)_1\right], \tag{3.27}$$

where $(\partial y/\partial x)_2$ is the slope of the wave at point 2, and $(\partial y/\partial x)_1$ is the slope of the wave at point 1. Note that both slopes are negative in our example and that $(\partial y/\partial x)_2$ is more negative than $(\partial y/\partial x)_1$. This result in a value for f which is also negative, i.e., the force f is a *restoring* force which acts in opposition to the displacement vector.

Since the segment is a differential segment, we may write the change in slope with longitudinal position between points 1 and 2 as

$$\frac{(\partial y/\partial x)_2 - (\partial y/\partial x)_1}{dx} = \frac{\partial^2 y}{\partial x^2},$$

so that Eq. (3.27) becomes

$$f = F\left(\frac{\partial^2 y}{\partial x^2}\right) dx. \qquad (3.28)$$

The term $\partial^2 y/\partial x^2$ is the *curvature*. Note that the restoring force is greatest where the curvature is greatest.

We may now apply Newton's second law to the segment,

$$f = ma \qquad (3.29)$$

where m, the mass of the differential segment, is given as $m = \mu\, dL$, where μ is the *mass per unit length* of string. Since a, the acceleration of the differential segment, is given as $\partial^2 y/\partial t^2$ at a constant value of x, Eq. (3.29) may be written as

$$f = (\mu\, dL)\frac{\partial^2 y}{\partial t^2}. \qquad (3.30)$$

Combination of Eq. (3.28) and Eq. (3.30) to eliminate f yields

$$F\frac{\partial^2 y}{\partial x^2}dx = (\mu\, dL)\frac{\partial^2 y}{\partial t^2}.$$

But, for small displacements, $dx \equiv dL$, so that

$$\frac{f}{\mu}\left(\frac{\partial^2 y}{\partial x^2}\right) = \frac{\partial^2 y}{\partial t^2}. \qquad (3.31)$$

At this stage it is convenient to arbitrarily define the constant F/μ. Equation then becomes

$$\frac{\partial^2 y}{\partial x^2} = \frac{1}{u^2}\left(\frac{\partial^2 y}{\partial t^2}\right) \qquad (3.32)$$

Equation (3.32) is the *general differential equation of wave motion in one dimension.*

The remaining problem is now to determine y as a function of x and t, that is, $y(x, t)$, such that the function satisfies Eq. (3.32). The resultant function $y(x, t)$ will then be a satisfactory equation for the traveling wave.

Since we know that y is a *periodic* function of both x and t, a possible general trial function might be

$$y = A \sin(ax + bt + c), \qquad (3.33)$$

where A, a, b, and c are constants. We must now evaluate the usefulness of this trial function by substituting it into Eq. (3.32). Thus.

$$\frac{\partial y}{\partial x} = Aa \cos (ax + bt + c).$$

$$\frac{\partial^2 y}{\partial x^2} = -Aa^2 \sin (ax + bt + c), \tag{3.34}$$

and

$$\frac{\partial y}{\partial t} = Ab \cos (ax + bt + c),$$

$$\frac{\partial^2 y}{\partial t^2} = -Aa^2 \sin (ax + bt + c) \tag{3.35}$$

Substitution of Eqs. (3.34) and (3.35) into Eq. (3.32) yields

$$u^2 a^2 = b^2,$$

or

$$b/a = \pm u. \tag{3.36}$$

Thus our trial solution is a satisfactory solution if, and only if, $b/a = \pm u$. Equation (3.33) may be rearranged to

$$y = A \sin a \left(x + \frac{b}{a} t + \frac{c}{a} \right) \tag{3.37}$$

Substitution of Eq. (3.36) into Eq. (3.37), and definition of $c/a = c'$, a new constant, yields

$$y = A \sin a (x \pm ut + c'). \tag{3.38}$$

Equation (3.38) is a satisfactory equation describing the displacement of a traveling wave as a function of position and time. We can easily deduce by inspection that the constant A is the amplitude, or maximum displacement, of the traveling wave, since the sine term must vary between +1 and −1, causing the displacement y to vary between $+A$ and $-A$.

It may be shown that Eq. (3.38) is applicable to all forms of smooth and uniform traveling waves, including sound waves and electromagnetic waves.

The Phase Velocity of a Traveling Wave

The *phase velocity* of a traveling wave is defined as $(\partial x/\partial t)_y$ and it may be represented graphically as the differential longitudinal motion dx of a point in the wave at some selected value of y in time interval dt Fig. (3.10). But, from Eq. (3.38), when y is constant,

$$A \sin a\,(x \pm ut + c') = \text{constant},$$

and since A and a are constants,

$$x \pm ut + c' = \text{constant}. \tag{3.39}$$

Taking the derivative of Eq. (3.39) with respect to t at constant y yields

$$\left(\frac{\partial x}{\partial t}\right) \pm u = 0,$$

or

$$\left(\frac{\partial x}{\partial t}\right)_y = \pm u = \text{phase velocity}.$$

The term u is thus the *magnitude* of the phase velocity,

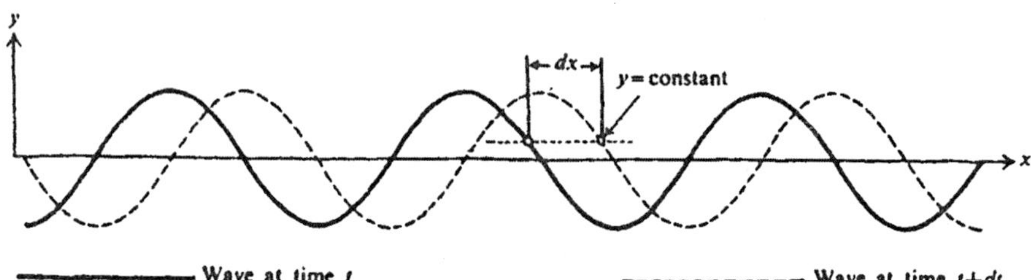

Fig. 3.10. The phase velocity of a wave is $(\partial x/\partial t)_y$.

The $\pm$ sign allows for propagation at the same speed in either direction. For a wave traveling in the $(+x)$- direction, that is, left to right, $(\partial x/\partial t)_y = +u$, which follows if a $(-)$ sign is used in front of ut in Eq. (3.38). Conversely, for a wave traveling in the $(-x)$-direction, $(\partial x/\partial t)_y = -u$, which follows if a $(+)$ sign is used in front of ut in Eq. (3.38).

From the original definition of u used in converting Eq. (3.31) into Eq. (3.32), the magnitude of the phase velocity may be expressed as

$$u = \sqrt{F/\mu}. \tag{3.40}$$

Thus, waves propagate more rapidly as the string becomes more tense or as it becomes lighter in mass.

Fixed Ends on a Stretched String : Boundary Conditions

Fixing both ends of a vibrating, stretched string at a *finite* length results in a wave motion which is quite different from the motion of the traveling wave in a taut string of *infinite* length. It is experimentally observed that when such a finitely confined string is plucked, evenly spaced *nodes* (points of zero displacement) and *antinodes* (points of maximum displacement) appear as shown for a typical mode in Fig. (3.11). The experimental conditions require, at the fixed ends be nodes and undergo no transverse displacement. But in addition we observe, depending on how the string is energized, an integral number of nodes (or none) *between* the two ends, with the string wagging transversely between any two nodes. Such a wave pattern is called a *standing wave.* The most interesting feature of the standing wave is that there is *no displacement at any time at the node positions.* Therefore, Eq. (3.38) for the traveling wave cannot directly apply to the standing wave since it allows for transverse displacement y at *any* value of x. We can think of the standing wave as being the additive wave produced by continued reflection of a traveling wave from fixed points at either end.

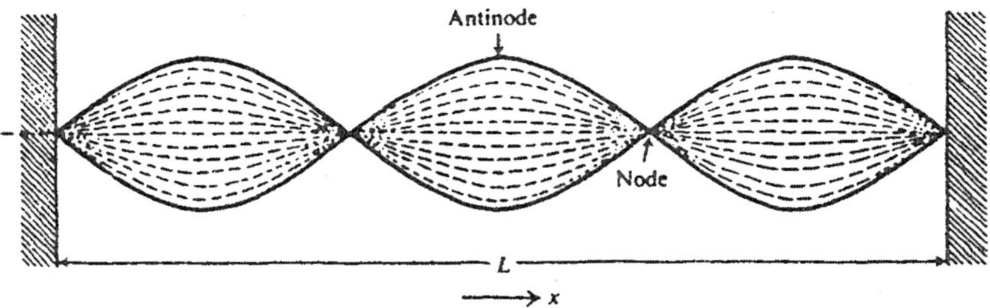

Fig. 3.11. A time exposure of a typical standing or stationary wave in a stretched string fixed at both ends. The loops wag transversely between the nodes.

Next we will attempt to develop a suitable function $y(x, t)$ to describe the variation of transverse displacement y with longitudinal position x and time t in a stretched string clamped on both ends at a distance L. In order to describe satisfactorily the behavior of a standing wave, the function $f(x, t)$ must meet two general requirements :

(a) It must satisfy the general differential equation of wave motion in one dimension,

$$u^2 \frac{\partial^2 y}{\partial x^2} = \frac{\partial^2 y}{\partial t^2}. \tag{3.32}$$

(b) It must satisfy the two *boundary conditions* imposed on the string by fixing the ends. These are

$$y = 0 \quad \text{when} \quad x = 0,$$

$$y = 0 \quad \text{when} \quad x = L.$$

We may use Eq. (3.38), which satisfy the general differential wave equation for a wave traveling in either direction in a taut string of infinite length. If the wave travels in the (+x)-direction (from left to right), the displacement y_+ is given as

$$y_+ = A \sin a\,(x - ut + c'), \tag{3.41}$$

whereas if the wave travels in the (−x)-direction (from right to left) the displacement y_- is given as

$$y_- = A \sin a\,(x - ut + c'), \tag{3.42}$$

The peculiar form of the standing wave results from the combination of the displacement of traveling waves which are constantly reflected from the fixed ends of the string, so that they travel in both directions simultaneously.

There is a well-known principle of wave theory, called the *principal of superposition,** which applies to the combination of linear wave equations such as Eqs. (3.41) and (3.42).

Consider, for example, two independent waves of the same amplitude but of different frequencies traveling in the same direction.

**Principle of superposition whenver two or more individual waves travel simultaneously in the same region, whether in the same or in different directions, the total displacement at any position is the sum of the displacements due to the individual waves.*

At certain points where the wave crests of the two waves coincide, the total displacement is twice the amplitude of each individual wave, and we say that the waves *reinforce* one another. At other points, where the wave crest of one wave happens to coincide with the wave trough of the second wave, the two waves cancel one another completely so that the total displacement is *zero*, and we say that the waves *destroy* one another. The general phenomenon of partial or complete reinforcement and destruction is referred to as interference.

For the taut string with fixed ends described above, the total transverse displacement y is given according to the principle of superposition as

$$y = y_+ + y_-, \tag{3.43}$$

$$y = A\,[\sin a\,(x - ut + c') + \sin a\,(x + ut + c')]$$

But, using the trigonometric identity

$$\sin\alpha + \sin\beta = 2\sin\left(\frac{\alpha+\beta}{2}\right)\cos\left(\frac{\alpha-\beta}{2}\right),$$

which holds for all values of α and β, we write Eq. (3.43) as

$$y = 2A\sin a(x + c')\cos aut,$$

or

$$y = A'\sin a(x + c')\cos aut, \tag{3.44}$$

where $A' = 2A$. Now lets determine what the constants a and c' must be in order for Eq. (3.44) to obey the two boundary conditions. At the left boundary of the string, $x = 0$ and $y = 0$. Substitution of zero for both x and y in Eq. (3.44) give

$$0 = A'\sin ac'\cos aut,$$

which must be valid for all values of t. If neither A' nor a is restricted to zero, then it is necessary that

$$\sin ac' = 0,$$

which is true whenever the argument of the sine is an integral multiple of π, that is, when

$$ac' = n'\pi, \tag{3.45}$$

where $n' = 0, \pm 1, \pm 2, \pm 3, \ldots, \pm \infty$. Substitution of c' from Eq. (3.45) back into Eq. (3.44) gives

$$y = A' \sin (ax + n'\pi) \cos aut. \tag{3.46}$$

At the right boundary of the string, $x = L$ and $y = 0$, and Eq. (3.46) becomes

$$y = A' \sin (aL + n'\pi) \cos aut,$$

which must again be valid for all values of t. If, once more, neither A' nor a is restricted to zero, then

$$\sin (aL + n'\pi) = 0,$$

which is true whenever

$$aL + n'\pi = n''\pi, \tag{3.47}$$

where $n'' = 0, \pm 1, \pm 2, \pm 3, \ldots, \pm \infty$. Rearrangement of Eq. (3.47) yields

$$a = \frac{(n'' - n')\pi}{L} = \frac{n\pi}{L}, \tag{3.48}$$

where n, which we define as a *positive* integer, may have the values $n = 0, 1, 2, 3, \ldots, \infty$. Substituting of a from Eq. (3.48) back into Eq. (3.46) yields

$$y = A' \sin \left(\pm \frac{n\pi x}{L} + n'\pi \right) \cos \left(\pm \frac{n\pi ut}{L} \right) \tag{3.49}$$

But the shift in the first term of the argument of the sine by $n'\pi$ results either in no change in the sine term itself in n' is an even integer, or in a change of sign if n is an odd integer. Either possibility may be accommodated if we write Eq. (3.49) as

$$y = \pm A' \sin \left(+ \frac{n\pi x}{L} \right) \cos \left(\pm \frac{n\pi ut}{L} \right)$$

Finally, since $\sin(-\propto) = -\sin \propto$ and $\cos(-\propto) = \cos \propto$, we can write Eq. (3.50) as

$$y = \pm A' \sin \left(\frac{n\pi x}{L} \right) \cos \left(\frac{n\pi ut}{L} \right) \tag{3.50}$$

which represents a *series* of satisfactory functions for the standing wave, including one positive function and one negative function for each value of n. For each of the positive functions, regardless of the value of n, the displacement y is a maximum and equal to A' whenever the sine term *and* the cosine term are each unity. Under these same conditions, the displacement is $-A'$ for each of the negative functions. In fact, for any given values of n, x, and t the displacements given by the positive and negative functions of Eq. (3.50) are always equal in magnitude but opposite in sign. Thus, it is apparent that any given negative function corresponding to a selected value of n is not really an independent function because it may be equally well represented by shifting the corresponding positive function by one-half period in time. Since the initial displacement of the wave at $t = 0$ is purely arbitrary, we'll choose to work only with the *positive* series of functions given by

$$y = A' \sin\left(\frac{n\pi x}{L}\right) \cos\left(\frac{n\pi ut}{L}\right) \qquad (3.51)$$

where $n = 0, 1, 2, 3, \ldots, \infty$. Let's now analyze the appearance of the string at some specific moment in time, which is equivalent to examining to snapshot photograph. With t constant, Eq. (3.51) becomes

$$y = A_t \sin\left(\frac{n\pi x}{L}\right) \qquad (3.52)$$

where $A_t = A' \cos(n\pi ut/L)$ = constant. Whenever nx/L is an integer, the argument of the sine term in Eq. (3.52) is an integral multiple of π, the sine term itself is zero, and the displacement y is zero. This is true regardless of the value of A_t, the amplitude of the fixed-time wave. Thus, *values of x for which nx/L is an integer define node positions,* which are positions of zero displacement at all times. Different node positions are generated for each value of n.

When $n = 0$,

$$nx/L = 0x/L,$$

which may be an integer (0) for all values of x. Zero displacement for all values of x represents a flat string with no energy of vibration.

When $n = 1$,

$$nx/L = x/L,$$

which may be an integer when $x = 0$ (integer = 0) or when $x = L$ (integer = 1). Since $x \leq L$, these are the only allowed values of x which result in nodes. The nodes occur at each end of the string, and the allowed waveform is shown in Fig. 3.12.

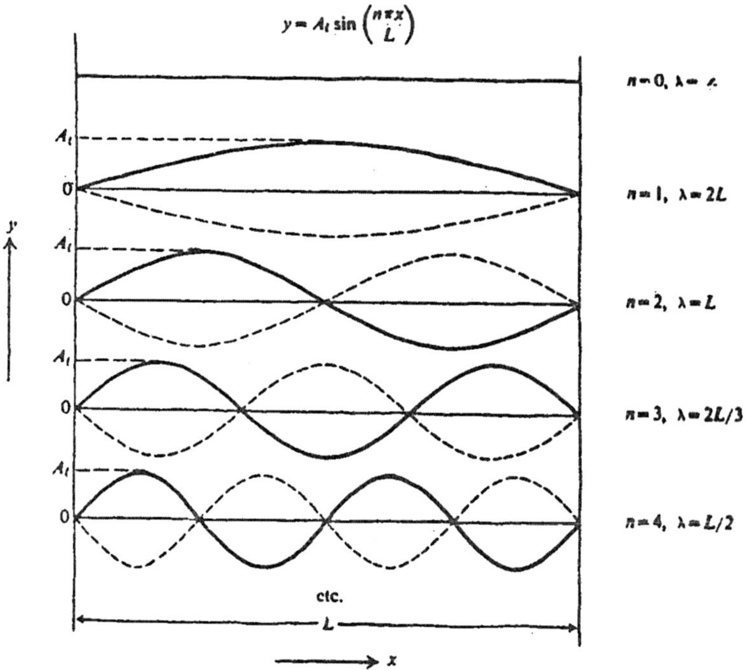

Fig. 3.12. Allowed modes of vibration for standing waves in a stretched string fixed at both ends. The dashed curves show the displacement of the string at a time which is one-half period later.

When $n = 2$,

$$nx/L = 2x/L,$$

which may be an integer when $x = 0$, $L/2$, or L. For $n = 2$, nodes thus occur at each end and in the middle of the string (Fig. 3.12).

When $n = 3$,

$$nx/L = 3x/L,$$

which may be an integer when $x = 0$, $L/3$, $2L/3$, or L. For $n = 3$, nodes then occur at each end, at $L/3$ and $2L/3$ (Fig. 3.12).

An examination of Fig. 3.12 indicates that *n is a direct measure of the number of antinodes in the standing wave.* Corresponding allowed *wavelengths* are also listed in Fig. 3.12. It is important to note that *only those wavelengths are allowed or persist for which L, the interval of confinement of the wave, is an integral multiple of one-half the wavelength.* All other wavelengths are lost through destructive interference, and the final waveform for the string must be a linear combination of any or all of the allowed modes. That is, the string may be vibrating in a waveform which represents a superposition of several or all of the allowed modes, depending on how the string is plucked or bowed.

We have shown that the imposition of boundary conditions on (or the confinement of) the traveling wave results in solutions which are restricted to a series of wavelengths related by integers. Only certain restricted forms of motion are allowed. It may be shown that the these equations are applicable to all forms of waves, including sound waves and electromagnetic waves.

THE STANDING WAVE AS A SYSTEM OF COUPLED LINEAR HARMONIC OSCILLATORS

In order to gain some insight into the time behavior of a particular particle in the string at some given value of x, we may hold x constant in Eq. (3.51) so that the displacement y is then a function of time only. Thus Eq. (3.51) may be written as

$$y = A_x \cos\left(\frac{n\pi ut}{L}\right) \tag{3.53}$$

where $A_x = A' \sin(n\pi x/L)$ = constant. The vibration of such a particle is shown in Fig. 3.13. We may consider the wave for which $n = 2$ as an example. The position of the string is shown at various successive times : t_1, t_2, t_3, t_4, t_5. Times t_1 and t_5 correspond to the positions of maximum and minimum displacement for *any* value of x. According to Eq. (3.53), a point in the string at any selected value of x will undergo simple harmonic motion, (the form of Eq. (3.53) is the same as that of Eq. (3.21), which defined simple harmonic motion).

If Eqs. (3.17) and (3.53) are to represent the same motion, the initial phase δ must be zero and the angular frequency must be given as

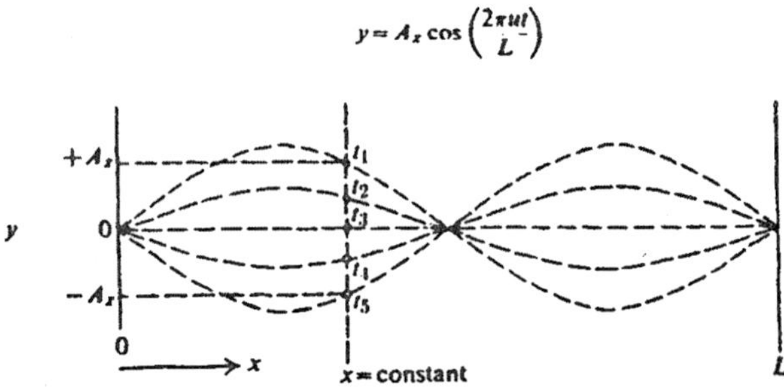

Fig. 4.13. The positions of a standing wave at successive times. Any point in the string undergoes simple harmonic motion.

$$\omega = n\pi u/L,$$

or, since $\omega = 2\pi\nu$,

$$2\pi\nu = n\nu u/L,$$

from which

$$\nu = nu/2L. \tag{3.54}$$

But in a stretched string, according to Eq. (3.40),

$$u = \sqrt{F/\mu},$$

so that combination of the above relationship with Eq. (3.54) gives

$$\nu = (n/2L)\sqrt{F/\mu}. \tag{3.55}$$

Equation (3.55) described the allowed frequencies in a stretched string which is fixed at both ends. Note that regardless of the position of x, all particles in the string oscillate on their y-axes at the same frequency. The entire string might thus be imagined as a longitudinal series of an infinite number of coupled simple harmonic oscillators, all vibrating at the same frequency on parallel transverse axes (Fig. 3.14).

A violin string is a good example of a fixed stretched string. The allowed frequencies are

$n = 1 : \nu_1 = \dfrac{1}{2L}\sqrt{\dfrac{F}{\mu}},$ fundamental (first harmonic),

$n = 2 : \nu_2 = \dfrac{2}{2L}\sqrt{\dfrac{F}{\mu}} = 2\nu_1,$ first overtone (second harmonic),

$n = 3 : \nu_3 = \dfrac{3}{2L}\sqrt{\dfrac{F}{\mu}} = 3\nu_1,$ second overtone (third harmonic), and so forth. The overtone frequencies are observed as integral multiples of the fundamental frequency. In practice, the displacement of a violin string usually results in a waveform which is a superposed form of many of the allowed frequencies with different amplitudes. Among other things, the exact tonal quality of a violin note depends on the particular combination of fundamental and overtones and on their relative intensities, which in turn depends on the way in which the string is bowed.

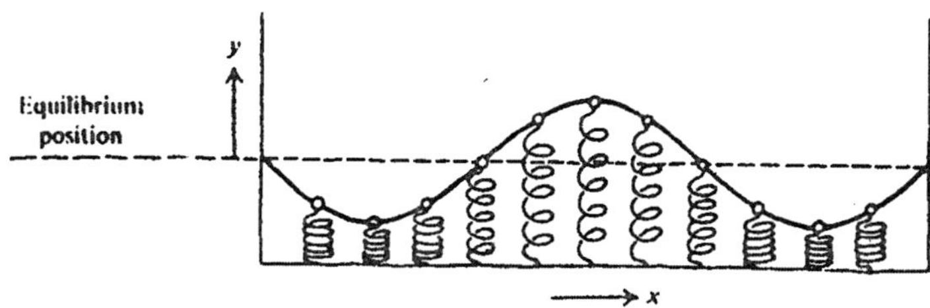

Fig. 4.14. The standing wave viewed as a series of coupled linear harmonic oscillators, all vibrating at the same frequencies but with different amplitudes on parallel transverse axes.

One may generally observe, however, that for standing waves having the same amplitudes at all the antinodes, larger values of n represent higher energies in the string. We have shown in the case of the simple linear harmonic oscillator (the mass on a spring) that the energy of the oscillator is directly proportional to the square of the amplitude (Eq. 3.22). This same relationship is true for each differential segment of the string in a standing wave, so that the energy per unit length of string or *energy density* at any point in the vibrating string is directly proportional to the square of the time-independent amplitude, A_x, at the point. That is,

$$\text{Energy density} \propto A_x^2.$$

Nodes thus represent positions of no energy and antinodes represent positions of maximum energy. It is clear, since A_x is a function of x, that the distribution of energy in the string between its end points is not uniform, but varies in such a way that the total energy density at any value of x remains constant with time. Furthermore, there are certain position of x where the energy density is zero and certain positions where the energy density is a maximum.

4

Two and Three Dimensional Waves

TWO-DIMENSIONAL WAVES

When a pebble is thrown into a pond, the resulting ripple wave travels in two dimensions. Similarly, waves travel in two dimensions in a stretched membrane in a drumhead. In fact, a vibrating drumhead represents a good two-dimensional analog of the one-dimensional stretched string fixed at both ends. We may calculate that only certain allowed standing two-dimensional waves may persist when the membrane is struck. Furthermore, two- dimensional nodes and antinodes occur. For the two-dimensional membrane the displacement D is a function of two coordinate positions x and y and time t. As for the stretched string fixed on both ends, only certain allowed types of *modes* of vibration persist, and the solutions for the allowed modes directly involve integers. The boundary conditions, which require that the displacement by zero along the entire periphery, lead to the simultaneous formation of standing waves in both the x- and y-directions. In each direction the allowed or persistent modes are those for which the distance across the membrane is an integral number of half-wavelengths. According to the principle of superposition, the total displacement D may be considered to be the sum of the displacements of the standing waves in the x- and y-directions.

It is important to recognize that the restriction of motion to discrete allowed modes of vibration or the membrane results from the requirement that the displacement function $d(x, y, t)$ simultaneously obey the *differential wave equation in two dimensions.*

$$\frac{\partial^2 D}{\partial x^2} + \frac{\partial^2 D}{\partial y^2} = \frac{1}{u^2}\left(\frac{\partial^2 D}{\partial t^2}\right) \tag{4.1}$$

and also obey the boundary condition, which requires that the entire periphery be a node. *If the two-dimensional waves were not confined, that is, if the surface of the membrane were infinite, nodes and antinodes would not arise.* Equation (4.1) is the two-dimensional analog of Eq. (3.22), which was the differential equation for the one-dimensional wave. The displacement D in two dimensions is analogous to the displacement y in one dimension.

THE GENERAL DIFFERENTIAL WAVE EQUATION IN THREE DIMENSIONS

Most of the problems of interest in quantum mechanics involve the confinement of particles having wave properties within three-dimensional enclosures. For example, the wave associated with the electron in the hydrogen atom must obey certain boundary conditions. It must also obey an appropriate differential wave equation.

The travel of an ordinary sound wave in air is a good example of a three-dimensional wave. If the sound wave is not confined, the wave motion may be represented as the emanation of spherical traveling waves which radiate outward in all directions for the source. If, however, we confine a three-dimensional sound wave within an enclosure such as a spherical cavity or a box (that is, apply boundary conditions), we cause interference through reflection of the sound waves from the boundaries, with the result that only certain allowed modes of vibration persist. That is, three-dimensional standing waves are produced which consist of only certain allowed combinations of nodes and antinodes. Displacement in this three-dimensional system may be represented in terms of the concentration or density of gas molecules. At those positions of x, y and z at which nodes occur in the standing wave, the concentration of gas molecules remains constant. At the antinode positions the extent of periodic fluctuation in gas density, that is, alternate compression and rarefaction, is maximum, and the sound intensity is greatest. At intermediate space positions between nodes and antinodes, the displacement, which is the difference between the concentration of gas molecules at a given position and the concentration of gas molecules at the node positions, periodically fluctuates with an amplitude somewhere between zero and the maximum amplitude given at the antinode positions.

In addition, at the antinode positions, the greatest number of gas molecules rush in and out of a given volume in a unit time so that the energy density associated with the wave motion is greatest at the antinodes. In fact, just as for the standing wave in one dimension, the energy density at a given position in the coordinate system is directly proportional to the square of the space-dependent amplitude at that point.

If we now define Φ as the displacement of a three- dimensional wave, where Φ is a function of the three coordinate positions x, y, and z and is also a function of time t, we may, write the *general differential equation of wave motion in three dimensions* as

$$\frac{\partial^2\Phi}{\partial x^2} + \frac{\partial^2\Phi}{\partial y^2} + \frac{\partial^2\Phi}{\partial z^2} = \frac{1}{u^2}\frac{\partial^2\Phi}{\partial t^2}. \tag{4.2}$$

We may then define the operator ∇^2 such that

$$\nabla^2 = \frac{\partial^2}{\partial x^2} + \frac{\partial^2}{\partial y^2} + \frac{\partial^2}{\partial z^2}, \tag{4.3}$$

so that Eq. (4.2) may be written more simply as

$$\nabla^2\Phi = \frac{1}{u^2}\frac{\partial^2\Phi}{\partial t^2}. \tag{4.4}$$

The operator ∇^2 ("del squared") is called the *Laplacian* operator.

Time-Average Displacement

Equation (4.4) applies to all wave motion, including electromagnetic waves, and expresses the displacement Φ as a function of both *position* and *time.* If our purpose is to locate the positions of the nodes and antinodes in *standing* waves produces by the imposition of boundary conditions, we are really interested in the *time-average* value of some displacement function at a particular point in space.

In the case of the equation for the one-dimensional standing wave (Eq. 3.51), in which the magnitude u of the phase velocity was constant and in which solutions were restricted through boundary conditions to a set of privileged constant values for ν, we were able to separate the complete displacement function into the product of a *space- dependent* part and a *time-dependent* part. It may be shown that when u is constant and when ν is restricted through boundary conditions to discrete constant values, we may make a similar separation in the equation for displacement of the *three-dimensional standing* wave.

By expressing $\Phi(x, y, z, t)$, the displacement function which must satisfy Eq. (4.4), as the product of $\psi(x, y, z)$, which depends only on position, multiplied by a periodic function which depends only on time, we may write

$$\Phi(x, y, z, t) = \psi(x, y, z)\, e^{-i\,2\pi\nu t},$$

where $i^2 = -1$ and ν is the frequency of oscillation of displacement in time at any space position. Or

$$\Phi = \psi e^{-i\,2\pi\nu t}. \tag{4.5}$$

THE TIME-INDEPENDENT CLASSICAL WAVE EQUATION

We can derive a time-independent wave equation from which the amplitude ψ can be evaluated by inserting Eq. (4.5),

$$\Phi = \psi e^{-i\,2\pi\nu t}.$$

into Eq. (4.4), the general differential wave equation,

$$\nabla^2\Phi = \frac{1}{u^2}\left(\frac{\partial^2\Phi}{\partial t^2}\right)$$

where u is constant, remembering that

$$\nabla^2 = \frac{\partial^2}{\partial x^2} + \frac{\partial^2}{\partial y^2} + \frac{\partial^2}{\partial z^2}.$$

Thus, assuming that ν is not a function of x, y, z, or t, that, is that ν is restricted to constant values in solutions of the general wave equation, and differentiating Eq. (4.5), we obtain

$$\frac{\partial\Phi}{\partial x} = \frac{\partial\psi}{\partial x} e^{-i\,2\pi\nu t}, \tag{4.6}$$

$$\frac{\partial^2\Phi}{\partial x^2} = \frac{\partial^2\psi}{\partial x^2} e^{-i\,2\pi\nu t}.$$

Similarly,

$$\frac{\partial^2\Phi}{\partial y^2} = \frac{\partial^2\psi}{\partial y^2} e^{-i\,2\pi\nu t}, \tag{4.7}$$

and

$$\frac{\partial^2\Phi}{\partial z^2} = \frac{\partial^2\psi}{\partial z^2} e^{-i\,2\pi\nu t}, \tag{4.8}$$

Also

$$\frac{\partial^2\Phi}{\partial z^2} = \frac{\partial^2\Phi}{\partial z^2} e^{-i\,2\pi\nu t},$$

$$\frac{\partial^2\Phi}{\partial t^2} = \psi\,(i\,2\pi\nu)^2\, e^{-i\,2\pi\nu t} = -\,\psi 4\pi^2\nu^2 e^{-i\,2\pi\nu t}. \tag{4.9}$$

Substitution of Eqs. (4.6), (4.7), (4.8), and (4.9) into Eq. (4.4) yields

$$\frac{\partial^2\psi}{\partial x^2} + \frac{\partial^2\psi}{\partial y^2} + \frac{\partial^2\psi}{\partial z^2} = \frac{-4\pi^2\nu^2\psi}{u^2}, \tag{4.10}$$

and substitution of the Laplacian operator symbol yields

$$\nabla^2\psi = \frac{-4\pi^2\nu^2}{u^2}\,\psi, \tag{4.11}$$

which may be rearranged to

$$\left(\frac{-u^2\nabla^2}{4\pi^2}\right)\psi = \nu^2\psi. \tag{4.12}$$

In solving Eq. (4.12) for three-dimensional systems in which waves are restricted or confined by boundary conditions we find, that solutions exist and may thus be obtained only for a privileged set of discrete values for ν^2, which correspond to the squares of the allowed frequencies for standing or stationary waves. The privileged ν^2-values are the eigenvalues of the operator $-u^2\nabla^2/4\pi^2$, and the corresponding ψ-functions which satisfy Eq. (4.12) are the eigenfunctions of the same operator. Thus, we'll usually find that a *series* of eigenfunctions satisfies the conditions of a particular problem, and that a *series* of eigenvalues results.

Furthermore, once we have evaluated the eigenfunctions, we may immediately write ψ^* for each of the solutions by substituting $(-i)$ for i wherever it appears so that we may write

$$\psi^*\psi = |\psi|^2 = |\Phi|^2, \tag{4.13}$$

from which we may determine where the energy density is localized for each of the modes.

It is also important to recall that the wave function is satisfactory *only* when ν is constant, that is, only when the system is in a stationary state. Thus Eq. (4.13) is a valid relationship only for *stationary states* in which ν is restricted to a constant. Since $u = \lambda\nu$, we can write Eq. (4.11) in a simpler form as follows :

$$\nabla^2\psi = -4\pi^2\psi/\lambda^2 \tag{4.14}$$

SECTION—II

GENERAL PRINCIPLES OF QUANTUM MECHANICS

5

The Quantum Hypothesis

The *Quantum* theory was *'born'* on December 14, 1900, in a meeting of the German Physical Society when Max Planck suggested that the spectral distribution of radiation emanating from hot objects may be accounted for *only* if one assumes that matter absorbs and emits energy in discrete bundles called *quanta.* Although the discrete division or *quantization* of *matter* into indivisible packets called atoms had been accepted since the beginning of the nineteenth century, the transmission and absorption of *energy* was still firmly believed to occur in the continuous fashion demanded by classical electromagnetic wave theory. The reluctance of scientists to accept Planck's proposal is understandable, since the acceptance of such a radical concept involved an inherent rejection of the fundamental framework of classical theory which had been developed to such a magnificent extent in the nineteenth century. Planck himself, was very disturbed with the significance of his own proposal and spent many years in a futile attempt to reconcile his quantum restriction with classical physics.

Einstein in 1905 provided the first major support for Planck's quantum hypothesis when he studied the theoretical aspects of radiation balance in a cavity and suggested, as a means of supporting his conclusions, that the photoelectric effect (the emission of electrons from the illuminated surfaces of certain metals) could be explained only in terms of quantization of light itself. He proposed that light is *propagated* in the form of *corpuscles* or packets in which the amount of energy is directed proportional to the frequency. These small packets or *quanta* of light were later to be called *photons* by G.N. Lewis in 1926.

In 1907 J.J. Thomson proposed a model for the atom in which discrete electrons were assumed to be distributed randomly within a cloud of positive charge such that the entire atom was neutral. Thomson's model has often been called the "plum pudding" model, the plums representing the electrons and the pudding representing the positive charge. This model, however, was short- lived. As a result of alpha-scattering experiments in 1911 Rutherford concluded that the positive charge must be concentrated in a very small, dense *nucleus* at the center of the atom and first introduced what we now know as the *nuclear atom.*

But there were serious problems in the stability of the Rutherford atom which could not be explained in terms of classical theory. In an attempt to solve these problems, the Danish physicist Niels Bohr in 1913 boldly proposed a *quantum* model for the hydrogen atom which provided remarkable agreement with the observed spectrum of hydrogen. Bohr's model was based on the postulate of quantized energy levels for the lone electron in the hydrogen atom, a postulate which had absolutely no justification in classical physics. But then again, neither were the earlier quantum postulates of Planck and Einstein justifiable in a classical sense. For the third time in 13 years scientists had encountered experimental observations which could be explained *only* in terms of quantized energy.

The period between 1913 and 1923 was spent largely in extending and patching the Bohr theory to fit the interpretation of more complex spectra, and the tools were becoming most unwieldy. However, a major breakthrough appeared in 1924 in the doctoral thesis of the French physicist Louis de Broglie, who proposed that the motion of electrons is intimately associated with *pilot waves.* In fact, he proposed a specific mathematical relationship between the momentum of a particle and the wavelength of its associated wave which supported Bohr's earlier model of the hydrogen atom.

Late in 1925 Werner Heisenberg, a German physicist, with the help of Max Born and Pascual Jordan, published his first paper on *matrix mechanics,* a technique which allows one to determine mathematically the possible transitions between different states of an atom and further enables one to determine the frequencies associated with such transitions. Matrix calculations led to results which were in closer agreement with experimental values than the most refined methods based on the older Bohr theory.

Almost simultaneously with the work of Heisenberg, Erwin Schrodinger, an Austrian physicist, developed the *wave mechanics*, in which he combined earlier aspects of classical wave theory with de Broglie's wave-particle relationship to postulate an entirely new technique for determining the properties of atomic systems. Schrodinger's methods are based on differential equations and are thus more easily comprehended by the average student than are the matrix calculations of Heisenberg.

A major refinement in quantum theory was made in 1929 by P. A. M. Dirac, a British physicist, who united the theory of relativity with quantum theory to produce the relativistic wave equation, which describes an atomic electron at velocities approaching the velocity of light. It also provides mathematically acceptable interpretations for such properties as intrinsic magnetic moments of electrons.

In order to understand where and how classical mechanics failed to provide satisfactory explanations for experimental observations, two developments must be considered in greater detail : Planck's quantum hypothesis of 1900 and Einstein's photoelectric theory of 1905. Planck and Einstein each found it necessary to postulate quantized energy levels in order to explain satisfactorily two apparently separate experimental phenomena involving the absorption of electromagnetic radiation.

WAVE PROPERTIES OF ELECTROMAGNETIC RADIATION

Visible light, heat rays, radio waves, and gamma radiation are different forms of the same fundamental radiation known as electromagnetic radiation. It is convenient to think of electromagnetic radiation as the propagation of a wave in which electrical and magnetic fields at right angles to one another periodically changes magnitude and direction (Fig. 5.1). Furthermore, electromagnetic waves have energy and are able to exchange this energy with matter. It may be shown that the energy of electromagnetic radiation at any point is proportional to the *square* of either the electrial or the magnetic vector.

The nature of a particular radiation is more precisely defined in terms of three variables : *wavelength, frequency,* and *speed of propagation.* The wavelength, λ, is the distance between two adjacent wavecrests or wavefronts and is often expressed in centimeters per cycle or simply in centimeters. The frequency, ν, represents the number of wavecrests or wavefronts which pass a given point in a unit time and is usually expressed in cycles/second or simply second^{-1}. Thus, it follows that the speed of propagation, *c*, of the wave may be obtained by multiplying the frequency by the wavelength :

$$c = \lambda\nu.$$

In terms of units : (cm/sec) = (cm/cycles)(cycles/sec).

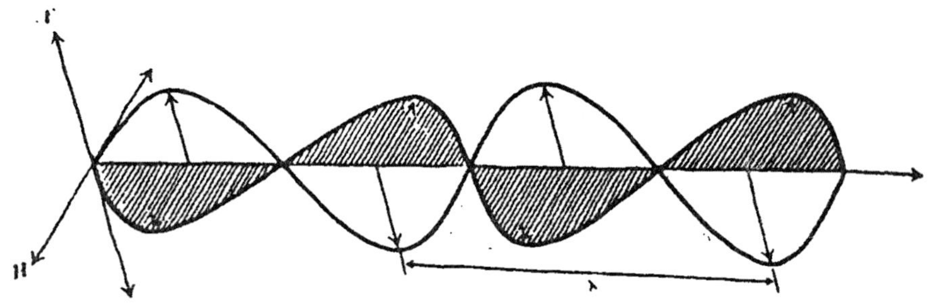

Fig. 5.1. the fluctuation of electrical (*E*) and magnetic (*H*) fields in a propagating electromagnetic wave (plane polarized).

For all forms of electromagnetic radiation the speed *c* in a vacuum is constant and is approximately equal to $3.0 \times 10^8 ms^{-1}$. Thus, electromagnetic radiation is classified conveniently either in terms of its wavelength, λ, or its frequency, ν, which is inversely proportional to the wavelength. The electromagnetic spectrum, shown in Fig. 5.2, includes radio waves several miles long at one end and very short, highly penetrating gamma rays at the other. Visible light constitutes only a very small portion of the total spectrum.

One of the most interesting and important aspects of electromagnetic radiation is the periodic electrical and magnetic disturbance produced at a given point in space as a wave passes by. An electron, for example, experiences a passing electromagnetic wave as a periodically fluctuating push and pull which causes it to oscillate at the same frequency as that of the wave. Any other charged particle would experience this same effect. Even more interestingly, the entire process may be reversed. That is, if a charged particle is forced to oscillate, electromagnetic radiation is generated. A floating cork subject to the effect of a ripple wave on the surface of a pond is a fair analogy to the electron subject to an electromagnetic wave, even though the ripple wave is quite complex in form. As the wave passes the position of the cork, the periodic fluctuation in the surface causes the cork to bob up and down at the same frequency as that of the wave. The forces on the cork operate primarily in a vertical or *transverse* direction, that is, perpendicular to the direction of wave propagation, just

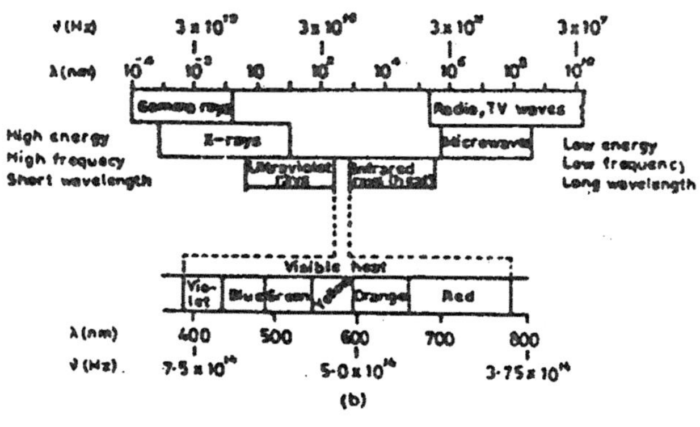

Fig. 5.2

as the electrical and magnetic forces in an electromagnetic wave are oriented perpendicularly to the direction of wave propagation. Conversely, if one begins with a smooth surface on the pond and mechanically forces the cork to oscillate up and down on the surface, a ripple wave is generated.

THE NATURE OF BLACKBODY RADIATION

All heated objects emit electromagnetic radiation. At relatively low temperatures a metal or ceramic rod emits mostly long-wavelength electromagnetic radiation known as infrared radiation, which is not visible but may be felt by the skin as heat. As the temperature is increased the predominant wavelength of the emitted radiation steadily decreases until a dull red visible radiation is observed at about 600°C. As the temperature is raised higher the color of the visible radiation gradually changes from red to yellow and eventually appears as a bright white incandescence at about 2000°C. At still higher temperatures significant ultraviolet radiation is also emitted.

The exact nature of the radiation, particularly the variation of intensity with wavelength, seems to depend to some extent on the solid being heated. For convenience, we may imagine a system in which the emitting solid is in complete equilibrium with the radiation so that the solid is hypothetically both a perfect absorber of radiation and also a perfect emitter. That is, it completely absorbs radiation of all wavelengths and in turn freely emits radiation of all wavelengths. Such a hypothetical solid is called a *blackbody radiator.* A perfectly dull black surface at least absorbs all of the *visible* radiation which impinges upon it. This is why it appears to be black. A perfect blackbody would totally absorb *all* electromagnetic radiation from gamma rays to radio waves.

The hypothetical "blackbody" may be approximated experimentally by constructing a hollow container (for example, a sphere) using a single pinhole to admit and release radiation. Radiation entering through the pinhold would have an extremely low probability of being reflected immediately back (Fig. 5.3). In effect it eventually would be completely absorbed. On the other hand, if we heated the sphere to incandescence, radiation leaving through the pinhole would also have had a very good chance to equilibrate with the incandescent surface before escaping. It would in effect represent *equilibrium radiation.*

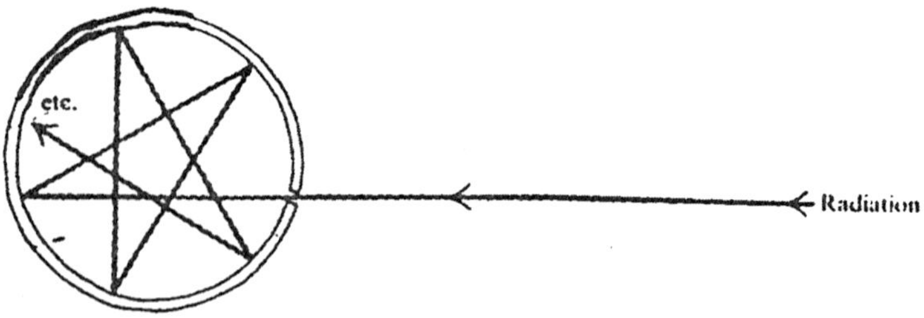

Fig. 5.3. A perfect blackbody radiator.

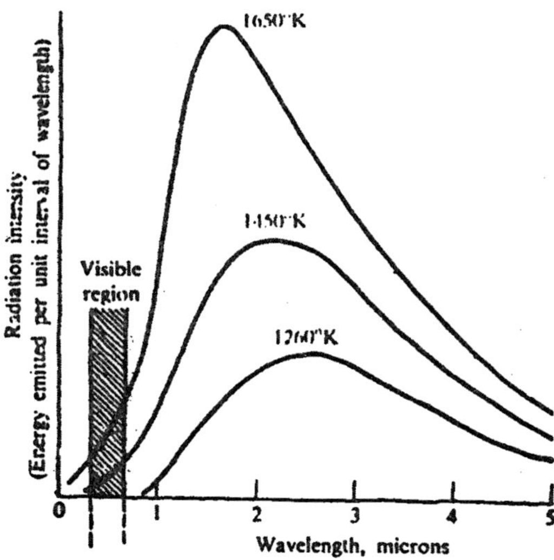

Fig. 5.4. Experimental curves of intensity versus wavelength for blackbody radiation.

It is found experimentally that the equilibrium radiation emitted from such a blackbody is identical for all solids and that the distribution of intensity of radiation as a function of wavelength is dependent only on the temperature of the incandescent solid. Typical experimental curves are shown in Fig. 5.4. Note that the wavelength of the most intense radiation (the peak) is lowered as the temperature is increased.

PERMITTED AND NON-PERMITTED VIBRATION

One of the major problems facing physicists at the end of the nineteenth century was that of explaining theoretically the nature of the exprimentally observed intensity versus wavelength curves for blackbody radiation.

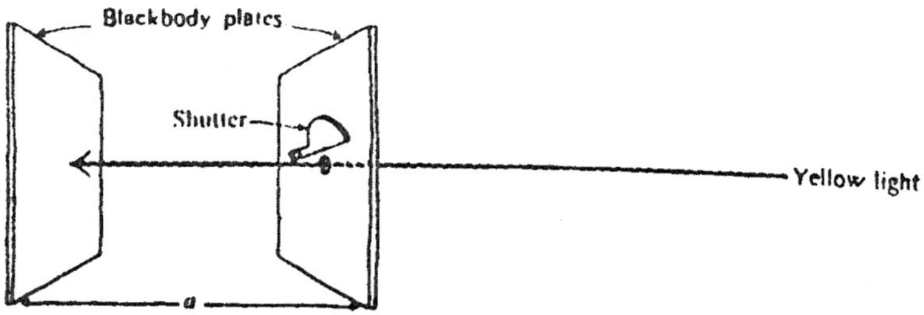

Fig. 5.5. A hypothetical one-dimensional blackbody cavity.

Classical physics attributes the origin of electromagnetic radiation in heated solids to the random vibrations and oscillations of charged nuclei and electrons. It is a well-known classical principle that charged particles emit electromagnetic radiation when accelerated. For example, the periodic to-and-fro acceleration of electrons in a radio transmitter antenna results in the emission of radio waves having the same frequency as the oscillating electrons. In a heated solid the myriad of random frequencies involved in the motion of the nuclei and electrons results in a random distribution of wavelengths in the emitted radiation.

However, in the case of the hypothetical blackbody, in which the radiation is forced to equilibrate within an enclosed cavity, classical wave theory dictates that only those wavelengths may persist (and may eventually be emitted) which may be reflected (or absorbed and immediately re-emitted) without destructive interference.

We may more easily understand the equilibration process in a blackbody cavity if we imagine the *one-dimensional* enclosure shown in Fig. 5.5. Such an imaginary device may be built by placing two parallel blackbody surfaces at an arbitrary separation distance, *a*. We may drill a pinhole in one plate and cover it with a moveable shutter. Now imagine that we open the shutter and allow electromagnetic radiation of *any* frequency distribution, for example, yellow light, to enter the cavity, and then immediately close the shutter. If the plates were considered to be merely parallel mirrors (which they are not), the radiation would be reflected back and forth in the single dimension. As we'll show more rigorously later, in such a reflecting case only those wavelengths are allowed and persist for which the distance between the plates is an integral multiple of one-half the wavelength, or for which

$$a = n\left(\frac{\lambda}{2}\right) \tag{5.2}$$

where n = an integer = 1, 2, 3, ..., ∞.

Such persistent waves are called *standing or stationary waves. All other wavelengths disappear through destructive interference.* Thus, if the plates were merely reflecting surfaces, the yellow light would in effect be "filtered" to leave only those wavelengths for which $\lambda = 2a/n$. If we were to now open the shutter after equilibrium had been reached, a distribution of such selected wavelengths would emerge through the pinhole.

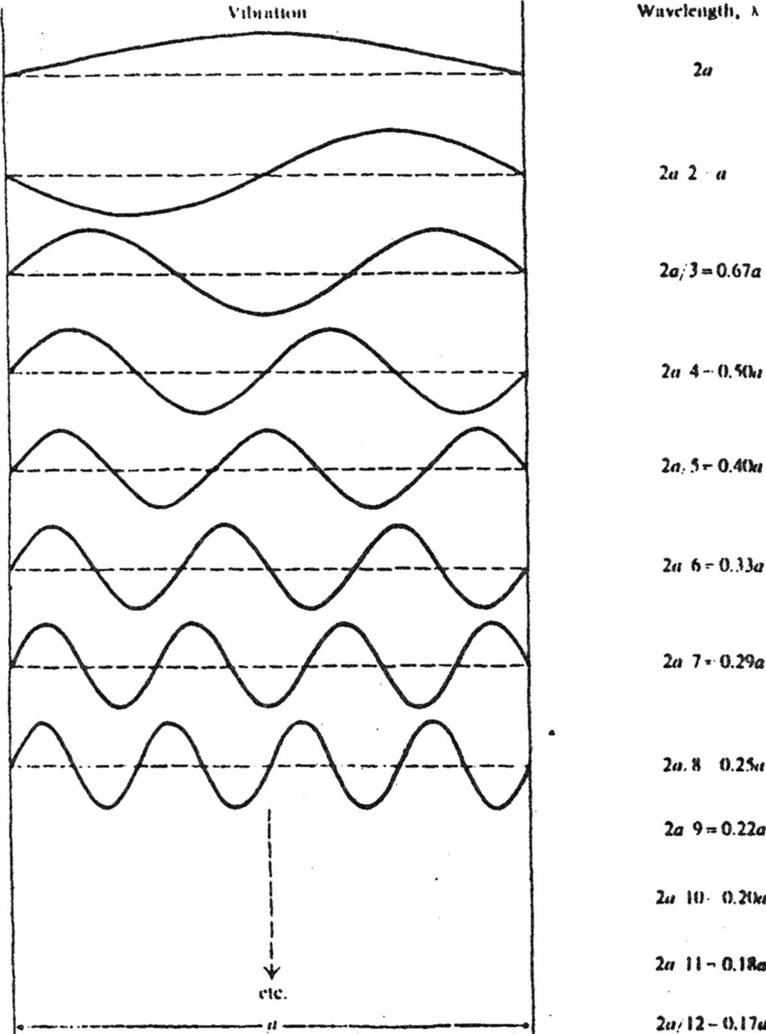

Fig. 5.6. Permitted vibrations in the one-dimensional blackbody cavity.

Blackbody surfaces, however, have the special characteristic of absorbing *all* wavelengths and emitting *all* wavelengths. For example, radiation of wavelength λ_1 might be absorbed by a blackbody surface and might be immediately re-emitted at the same wavelength λ_1, or it might be stored briefly and then be re- emitted by the same or an adjacent oscillator without loss of phase relationship. In either case the process is equivalent to reflection and would be expected to be a common occurrence, since a particular oscillator in the solid which absorbs radiation of wavelength λ_1 also emits radiation of wavelength $\lambda 1$, might be absorbed, stored briefly as energy in the blackbody surface, and then might be re-emitted at a *different* wavelength λ_2, *which may be higher or lower than* λ_1, since the blackbody is capable of emitting radiation of all wavelengths. This latter process is equivalent to an *exchange* of frequencies. The eventual result of a very large number of "reflections" and wavelength exchanges would be an equilibrium distribution which contains *all* of the allowed or persistent wavelengths.

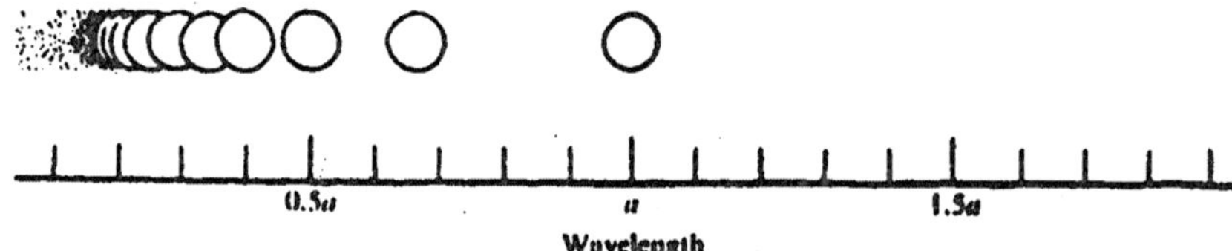

Fig. 5.7. Permitted vibrations as a function of wavelength in the one-dimensional blackbody cavity.

Thus, if yellow light (or any other light for that matter) were introduced into the hypothetical one-dimensional blackbody cavity, it would be converted through eventual equilibration to *all* of the allowed wavelengths. The cavity would contain all sorts of radiation from γ-rays increasing in wavelength to the maximum allowed wavelength, $2a$ (in which $n = 1$).

It is now apparent that the introduction of *any* electromagnetic radiation into the cavity or the generation of *any* electromagnetic radiation within the cavity through incandescence would lead to the same eventual classical distribution of wavelengths, which at equilibrium would contain all vibrations for which

$$\lambda = \frac{2a}{n}, \; n = 1, 2, 3, \ldots, \infty.$$

All other vibrations would disappear through destructive interference.

For the one-dimensional model, the persistent or "allowed" vibrations are graphically depicted in Fig. 5.6. The allowed vibrations as a function of a wavelength are depicted in Fig. 2.7. Observe that the points representing allowed vibrations merge together as λ approaches zero. That is,. the number of allowed vibrations approaches infinity as the wavelength of the vibration approaches zero.

THE PHOTOELECTRIC EFFECT

Hertz first noted in 1887 that electrons are emitted if light is allowed to fall on a clean metal place in a vacuum. A simple device for illustrating this phenomenon, which is called the *photoelectric effect,* is shown in Fig. 5.8. A metal lace, from which electrons are to be photoemitted, is placed in an evacuated tube and voltage is applied between it and another more positive electrode so that a current may be detected in the galvanometer, *G*, whenever electrons are emitted. Not all metals readily show the photoelectric effect. For most metals, ultraviolet radiation is necessary, but metals which hold their electrons more loosely, such as the alkali metals, potassium and cesium release electrons easily when bombarded by visible light.

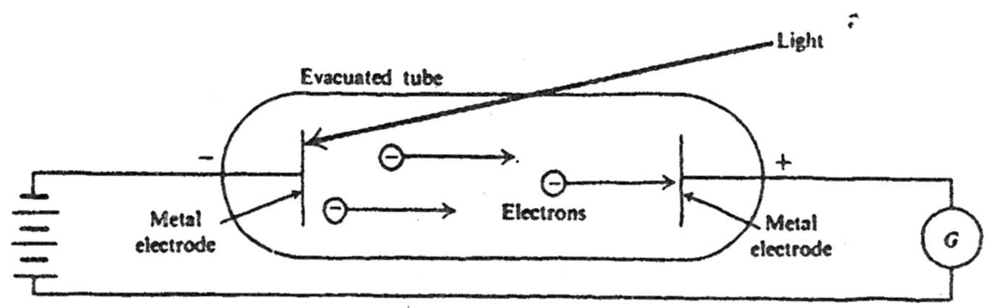

Fig. 5.8. The photoelectric effect.

The following three observations are impossible to explain in terms of classical theory :

1. Although the *number* of electrons emitted is dependent on the intensity of radiation, the *energy* of the average emitted electron is *not dependent* on the intensity of incident radiation. According to classical physics, if radiation were absorbed continuously, electrons, should be able to absorb more energy from a more intense beam before escaping.

2. The maximum energy of escaping electrons is shown to depend on the *frequency* of the incident radiation. In fact, there is observed to be a *threshold frequency*, ν_0, different for each metal, below which photoemission does not take place.

3. There is no apparent lag between the time the light strikes the plate and the time electrons are emitted. Classical physics requires a time during which the evenly distributed, continuous radiant energy accumulates to a level sufficient for the electron to overcome surface forces in the metal and escape.

While explaining the observed data, Einstein proposed in 1905 that *electromagnetic radiation itself is quantized*, just as Planck had previously postulated that energy levels in a blackbody oscillator are quantized. Einstein visualized the elementary act of photoemission as resulting from the collision of one corpuscle or quantum of electromagnetic radiation (later to be called a *photon*) with one electron in the surface of the metal. He further postulated that *each photon has an energy hν*. He proposed in effect that Planck's quantized energy levels in the blackbody oscillator are not really fundamental properties of the oscillator alone but result from *quantized energy levels in electromagnetic radiation itself.*

There is a minimum energy, ϕ, called the work function, which must be expended in order to remove an electron from the surface of a metal. The value of ϕ varies from metal to metal and depends on the tenacity with which a given metal holds its electrons. The work function for alkali metals is relatively low compared to that for transition metals. Assume that one photon of energy $h\nu$ is absorbed by one electron in the surface of the metal :

1. If $h\nu < \phi$, the electron cannot be emitted, since it does not acquire sufficient energy to escape.

2. If $h\nu > \phi$, the electron may escape and the excess energy appears as kinetic energy in the emitted electron. That is,

$$h\nu = \phi + \text{kinetic energy}. \tag{5.3}$$

Since the kinetic energy of an electron of mass m and velocity v is given as $\frac{1}{2}mv^2$,

$$h\nu = \phi + \tfrac{1}{2}mv^2, \tag{5.4}$$

or

$$\frac{1}{2}mv^2 = h\nu - \phi. \qquad (5.5)$$

Equation (5.5) predicts that the kinetic energy of an emitted electron should depend only on ν, the frequency of the incident radiation. An increase in *intensity* of radiation only increases the *number* of electrons emitted, not the kinetic energy. Furthermore, if ν is too low, that is, if $h\nu < \phi$, the electron cannot escape. Apparently, then, the threshold frequency ν_0 is given by

$$h\nu_0 = \phi$$

or

$$\nu_0 = \phi/h. \qquad (5.6)$$

A helpful graphical picture is provided if we note that Eq. (5.5) is the equation of a straight line where $\frac{1}{2}m_c\nu^2$ is the ordinate and ν is the abscissa. Then h is the slope and $-\phi$ is the ordinate intercept. The linear plot is shown in Fig. 4.9.

Equation (5.5) also predicts that all metals should show the same slope on a plot of kinetic energy versus frequency of incident radiation. It was not until 1916 that Millikan experimentally verified Einstein's prediction by making careful measurements conducted in an attempt to *disprove* Einstein's photoelectric theory.

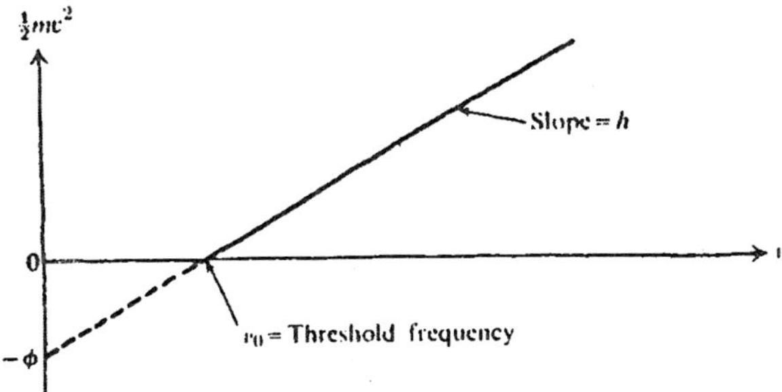

Fig. 5.9. Kinetic energy of a photoemitted electron as a function of the frequency of the incident radiation.

Example 5.1. A spectrum consists of violet light of frequency $7.31 \times 10^{14}s^{-1}$ and of red light of frequency $4.57 \times 10^{14}s^{-1}$. Calculate the energy of an individual photon in each of these two colours of light in joules ($h = 6.626 \times 10^{-34}Js$).

Solution :

(violet light) $E = h\nu$

$= (6.626 \times 10^{-34} Js)(7.31 \times 10^{14} s^{-1})$

$= 4.85 \times 10^{-19} J$

(red light) $= h\nu$

$= (6.626 \times 10^{34} Js)(4.57 \times 10^{14} s^{-1})$

$= 3.03 \times 10^{-19} J$

Example 5.2. The maximum kinetic energy of the photoelectrons emitted from a metal is 1.03×10^{-19} J when light that has a 656 nm wavelength shines on the surface. Determine the threshold frequency ν_0 for this metal.

Solution :

Step-I : We have

$\lambda = nm = 656 \times 10^{-9} m$

$c = 3.00 \times 10^{8} ms^{-1}$

Thus

$$\nu = \frac{c}{\lambda} = \frac{3.00 \times 10^{8} ms^{-1}}{(6.56 \times 10^{-9} m)} = 4.57 \times 10^{14} s^{-1}$$

Step-II : Rearranging the expression $h\nu = h\nu_0 + \frac{1}{2} mv^2$, we get

$$\nu_0 = \frac{h\nu - \frac{1}{2} mv^2}{h}$$

$$= \frac{(6.626 \times 10^{-34} Js)(4.57 \times 10^{14} s^{-1}) - (1.03 \times 10^{-19} J)}{6.626 \times 10^{-34} Js}$$

$= 3.02 \times 10^{14} s^{-1}$

Hence a minimum frequency of 3.02×10^{14} Hz is required to evoke the photoelectric effect for this metal.

Example 5.3. Calculate the work function and threshold wavelength for Na if $v_0 = 4.39 \times 10^{14} s^{-1}$.

Solution : We have

$$h\nu_0 = h\nu_0 (6.626 \times 10^{-34} Js)(4.39 \times 10^{14} s^{-1})$$

$$= 2.91 \times 10^{-19} J$$

$$\lambda_0 = \frac{c}{u_0} = \frac{3.00 \times 10^8 s^{-1}}{4.39 \times 10^{14} s^{-1}} = 0.683 \times 10^{-6} m$$

$$= 683 \times 10^{-9} m$$

$$= 683 nm$$

After all, what now is electromagnetic radiation? Is it wavelike or is it corpuscular? It seems impossible that it can be both simultaneously. Yet, we must consider light to be wavelike if we are to explain diffraction and interference and we must consider light to be corpuscular if we are to explain the photoelectric effect. The general concept of *wave-particle dualism* made its debut with Einstein's photoelectric hypothesis in 1905. There seemed to be no alternative choice. For a second time in five years scientists were forced to resort to quantization of energy experimental observations.

6

The Bohr Model and Matter Waves

The nuclear model of atom (1912) presented many problems to the classical physicist. Alpha-particle scattering experiments (1911) led Rutherford to propose that the atom as composed mostly of empty space with electrons revolving at relatively great distances around a small, very dense, positive nucleus.

Classical electromagnetic theory requires that an accelerated charged particle (one which is changing either speed or direction or both) must emit electromagnetic radiation. The planetary electron of the Rutherford model, in moving in a circular orbit, undergoes constant acceleration toward the nucleus and thus should continuously emit radiation. But if the electron emit radiation, its energy would decrease, its velocity would drop and it would begin to spiral toward the nucleus, constantly losing energy as its orbital radius decreases. The ultimate result would be *complete collapse of the electrons* into the nucleus! Classical calculations show that the entire collapse should occur in one hundred-millionth of a second (10^{-9} sec)! Of course, such a collapse does not occur. Neither do we observe the continuous emission of radiation. Rather, excited atoms emit radiation only at selected discrete wavelengths in what is called a *line spectrum.*

LINE SPECTRA OF ATOMS

It is well known that salts of certain metals, when heated to high temperatures impart characteristic colours to the emitted light. Thus, sodium salts impart a yellow colour to a flame, potassium salts impart a violet colour, and barium salts show a light green colour. In addition, when an electrical discharge is passed through a gas at a sufficiently low pressure the gas emits a characteristically coloured glow discharge. The orange-red discharge of neon, for example, is commonly seen in commercial neon signs. Mercury vapour emits a greenish-blue glow discharge, and hydrogen emits a violet-blue discharge.

We may more completely characterize such flame colours or glow discharges by dispersing the emitted light through a prism so that the various wavelengths are spread out into a spectral pattern. An appropriate experimental arrangement is shown in Fig. 6.1. The slits serve to collimate the beam into a narrow hand of light which enters the prism. Were visible white light used as the source, the narrow band would be dispersed into a continuous spectrum from red light at the long-wavelength end to violet light at the short-wavelength end. When, however, the flame coloured by a metal salt or the glow discharge of a gas is used as the source, the spectrum consists of a series of characteristic sharp lines, which reproducibly appear at the same wavelengths for a given atom. Foucault, for example, accurately identified in 1848 the principal sharp line in the emission spectrum of sodium which is responsible for its yellow colour in a flame. Further work by Kirchhoff and Bunsen in 1861 identified the principal sodium line to be actually a pair of closely spaced lines (a doublet) at a wavelength of about 5890 Å.

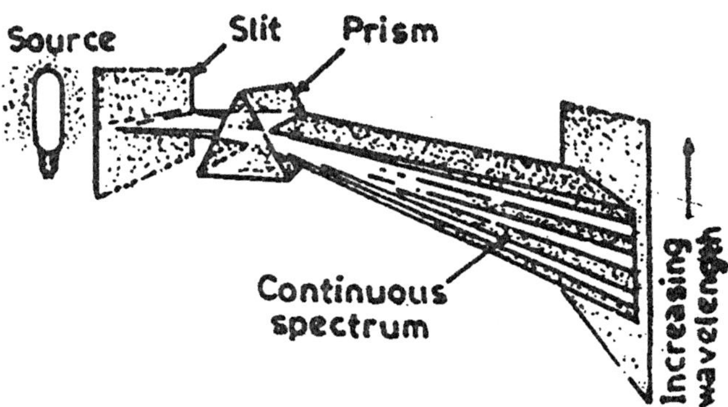

Fig. 6.1. Dispersion of light by a prism.

Because of the characteristic association of discrete line spectra with certain elements and because of the sharpness of the lines, it soon became convenient to use emission spectroscopy as a tool for the analysis of complex mixtures of elements.

The Hydrogen Spectrum

Balmer in 1885 carefully examined the series of about 35 resolvable lines appearing in the visible and near-ultraviolet regions of the emission spectrum of atomic hydrogen (Fig. 6.2). It is instructive to note that the lines are spaced more and more closely together as their wavelengths decrease until a continuum finally appears at 3646.6 Å.

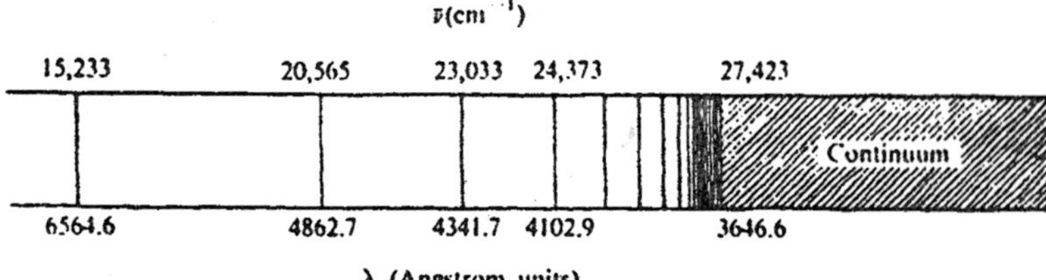

Fig. 6.2. The Balmer series for hydrogen.

Balmer found that the wavelengths in vacuo of the spectral lines may be mathematically related in a purely empirical way which may be expressed as

$$\bar{\nu} \equiv \frac{1}{\lambda} = R_H\left(\frac{1}{2^2} - \frac{1}{n^2}\right) \tag{6.1}$$

where $\bar{\nu}$ is called the *wave number* and is defined as $1/\lambda$, R_H is a constant called the Rydberg constant, and n may have the integral values 3, 4, 5, 6, ..., ∞. The Rydberg constant is determined experimentally to be 109,677.60 ± 0.03 cm^{-1}.

In 1906 Lyman discovered another similar spectral series for hydrogen in the ultraviolet region. Paschen detected a third series in the infrared region in 1909. It was soon found that all of the spectral emission lines for hydrogen can be expressed by a more generalized form of Eq. (6.1), called that Rydberg equation.

$$\nu = \frac{1}{\lambda} = R_H\left(\frac{1}{n_1^2} - \frac{1}{n_2^2}\right) \tag{6.2}$$

where $R_H = 1.097 \times 10^7 m^{-1}$ is known as the **Rydberg constant,** λ is the wavelength and n_1 and n_2 are positive integers (quantum numbers); n_1 is smaller than n_2.

It is important to note that the relationship given by Eq. (6.2) was deduced entirely from empirical data and until 1913 there was no theoretical explanation for the origin of discrete spectral lines.

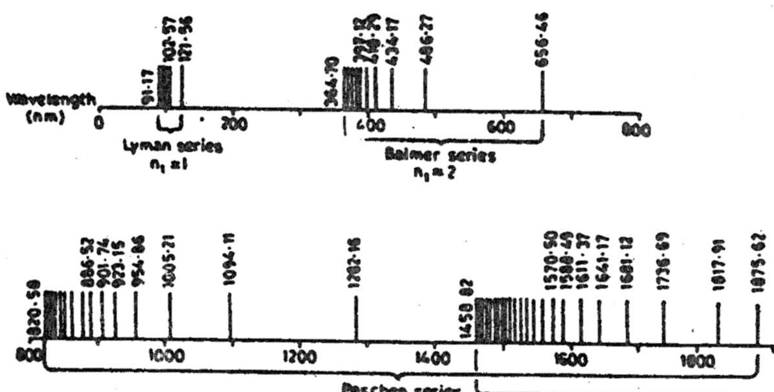

Fig. 6.3. The spectrum of hydrogen atom showing the series limits.

The five spectral series of hydrogen atom along with the values of n_1 and n_2 are given in Table 6.1.

TABLE 6.1
Spectral Series of Hydrogen Atom

Spectral series	n_1	n_2	*Spectral region*
Lyman	1	2,3,4	ultraviolet
Balmer	2	3,4,5	visible
Paschen	3	4,5,6	near infrared
Brackett	4	5,6,7	infrared
Pfund	5	6,7	far infrared

The wavelengths of the lines of the five series and their limits are given in Fig. 6.3. Lyman series appears in ultraviolet region, Balmer series in the visible region, while the other three (Paschen, Brackett and Pfund) appear in the infrared region.

Example 6.1. Calculate the wavelength of radiation corresponding; $n_1 = 2$ and $n_2 = 3$ in the hydrogen spectrum;

$$R_H = 1.097 \times 10^7 m^{-1}.$$

Solution : Substituting the values in the expression

$$\bar{\nu} = \frac{1}{\lambda} = R_H\left(\frac{1}{n_1^2} - \frac{1}{n_2^2}\right)$$

we have

$$\frac{1}{\lambda} = (1.097 \times 10^7 m^{-1})\left(\frac{1}{2^2} - \frac{1}{3^2}\right) = \frac{1.097 \times 10^7 \times 5}{36} m$$

or $\qquad \lambda = 6.56 \times 10^{-7} m$

BOHR'S MODEL FOR THE HYDROGEN ATOM

In 1913, Niels Bohr postulated a model for the hydrogen atom with which he attempted to solve several of the perplexing conflicts between classical theory and experiment.

The postulates for Bohr's model of the hydrogen atom are :

1. The single electron in the hydrogen atom moves about the nucleus in circular orbits in which the electrostatic attraction between the negatively charged electron and the positively charged nucleus is exactly balanced by the centrifugal force due to the velocity of the electron in its orbit.

2. While in a give orbit, the energy of the electron is constant. Furthermore, the electron remains in a given orbit unless a quantum of energy of exactly the right amount is absorbed or emitted.

This postulate, often called the *stationary-state assumption,* is contrary to the ordinary laws of electrodynamics, since the accelerating electron should constantly lose energy.

3. Only certain orbits are allowed. Such orbits are restricted to those for which the angular momentum L of the electron is an integral multiple of $h/2\pi$ or

$$L = mvr = nh/2\pi, \tag{6.3}$$

where m is the mass of the electron, v is its linear velocity, r is the radius of the orbit, and n may assume the values 1, 2, 3, ..., ∞.

4. When an electron makes a transition from an initial stationary energy state E_{n_2} to a lower final stationary energy state E_{n_1} the energy difference is *emitted* as a single photon such that

$$E_{n_2} - E_{n_1} = hv, \tag{6.4}$$

where ν is the frequency of the corresponding spectral line. Conversely, if energy is absorbed by an electron, the same relationship must apply. That is, the wavelength of the absorbed photon is given by Eq. (6.4). (This postulate is in accord with Einstein quantum hypothesis for electromagnetic radiation.)

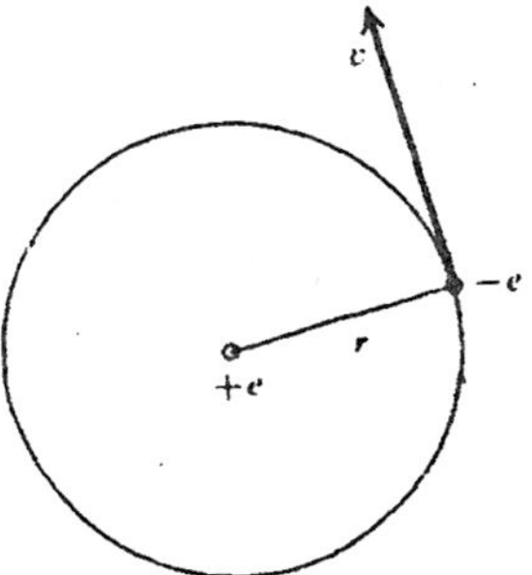

Fig. 6.4. Orbital motion of the electron about the nucleus.

With the help of the above postulates, the problem can be tackled mathematically as follows.

Consider a single electron of charge $-e$ revolving with a linear velocity v in a stable circular orbit of radius r about the single proton (charge = $+e$), which is the nucleus of hydrogen. The relationship is illustrated in Fig. (6.4). According to the first postulate, the force of electrostatic attraction F_A between the electron and the nucleus must be equal to the centrifugal force F_C which tends to separate them :

$$F_A \text{ (force of attraction)} = F_C \text{ (centrifugal force).} \qquad (6.5)$$

Coulomb's law states that the force acting between two charges q_1 and q_2 at a separation r is given by

$$F \propto \frac{q_1 q_2}{r^2}.$$

Usually, absolute values for q_1 and q_2 are used, in which case like charges result in a force of repulsion while unlike charges result in a force of attraction. Ordinarily a proportionality is converted to an equation by inserting a constant on one side. In the cgs system we assign unit value to the constant by defining the electrostatic unit of charges (esu or statcoulomb) as that quantity of electricity which, when placed in a vacuum 1 cm from an equal charge, will be acted on by a force of 1 dyne. Thus, in a vacuum, by definition,*

$$F = \frac{q_1 q_2}{r^2}, \qquad (6.6)$$

where F is expressed in dynes (g-cm-sec^{-2}), r is expressed in centimeters, and q is expressed in esu or statcoulombs (g$^{1/2}$-cm$^{3/2}$-sec^{-1}). The charge on the electron, e, is 4.80298×10^{-10} esu. Of course, for hydrogen this is also equal to the charge on the nucleus.

Noting that centrifugal force, F_c is given by mv^2/r, where m is the mass of the electron, we may write Eq. (6.5) as

$$\frac{q_{\text{nucleus}}\, q_{\text{electron}}}{r^2} = \frac{mv^2}{r},$$

or, using an absolute value of e for the charge on both the electron and the nucleus and noting that the force is an attractive force, as

$$e^2/r = mv^2. \qquad (6.7)$$

The total energy E of the electron in its orbit is given as the sum of its kinetic energy T and its potential energy V :

$$E = T + V. \qquad (6.8)$$

The kinetic energy T is given as $\frac{1}{2}mv^2$. The potential energy, however, is a relative value which depends on the state we define as having zero potential energy. *It is convenient to arbitrarily define the system to have a potential energy of zero when the electron is infinitely removed from the nucleus.* As the electron is brought from a distance ∞ to a distance r, the system could be made to do work on the surrounding due to the attractive force between the nucleus and the electron. Thus, the potential energy decreases as r decreases. The total amount of work the electron does in moving from ∞ to a distance r is thus equal to the potential energy V at a distance r and is given by

$$V = \int_{\infty}^{r} F_A\, dr = \int_{\infty}^{r} \frac{e^2}{r^2}\, dr = e^2 \int_{\infty}^{r} \frac{dr}{r^2},$$

*In the SI system, Coulomb's law is written as $F = (1/4\pi\varepsilon_0)q_1q_2/r^2$, where F is expressed in newtons, r in meters, q_1 and q_2 in coulombs (ampere-seconds), and ε_0 is called the permittivity constant, which in a vacuum is equal to 8.85415×10^{-12} coulomb^2-newton^{-1}-meter^{-2}.

or $$V = -e^2/r.$$

Substitution of the values for T and V into Eq. (6.8) yields

$$E = mv^2/2 - e^2/r. \tag{6.9}$$

We may now substitute Eq. (6.7) into Eq. (6.9) to obtain

$$E = -mv^2/2 = -e^2/2r. \tag{6.10}$$

Substitution of the quantum restriction equation,

$$mvr = nh/2\pi, \tag{6.3'}$$

into Eq. (6.10) yields

$$v = 2\pi e^2/nh, \tag{6.11}$$

where $n = 1, 2, 3, 4, \ldots, \infty$.

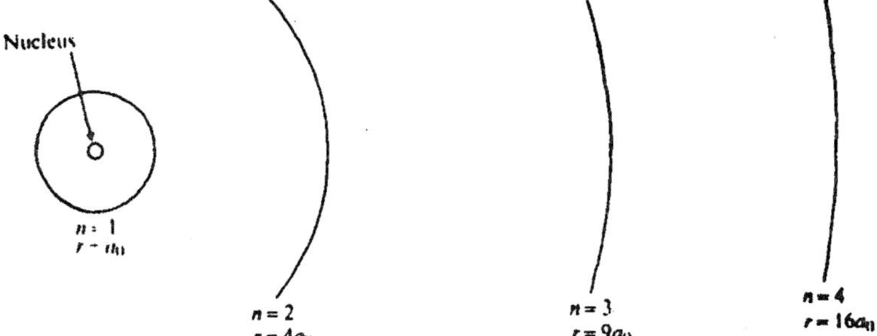

Fig. 6.5. Permitted radii for the first four Bohr orbits of hydrogen (a_0 = the radius of the first Bohr orbit).

The integer n is called the *principal quantum number*. Substitution of various values of n into Eq. (6.11) generates the discrete linear velocities of the electron in the allowed Bohr orbits. Allowed energies in each of the stationary states or orbits may be generated by squaring Eq. (6.11),

$$v^2 = 4\pi^2 e^4/n^2h^2,$$

and combining with Eq. (6.10) to eliminate v^2. Thus

$$E = -2\pi^2 me^4/n^2h^2, \tag{6.12}$$

where $n = 1, 2, 3, \ldots, \infty$.

It is also convenient to combine Eq. (6.12) with Eq. (6.10) to eliminate E :

$$e^2/2r = 2\pi^2 me^4/n^2h^2,$$

from which the allowed radii for the Bohr orbits are given as

$$r = n^2h^2/4\pi^2 me^2, \qquad (6.13)$$

where $n = 1, 2, 3, ..., \infty$.

Equations (6.12) and (6.13) can be further simplified by grouping constants. Thus

$$E = -R(1/n^2), \qquad (6.14)$$

where

$$R = 2\pi^2 me^4/h^2, \qquad (6.15)$$

and

$$r = a_0 n^2, \qquad (6.16)$$

where

$$a_0 = h^2/4\pi^2 me^2. \qquad (6.17)$$

The permitted orbital radii for the Bohr model obtained by inserting successive integers into Eq. (6.16), are

$$a_0, 4a_0, 9a_0, 16a_0, 25a_0, ...$$

Relative orbit sizes are shown in Fig. 6.5. When $n = 1$ in Eq. (6.16), a_0 is shown to be the radius of the first Bohr orbit for hydrogen. The calculated value of a_0 from Eq. (6.17) is 0.529167 Å, which is in good agreement with the estimated size of the hydrogen atom in the closely packed solid.

The real triumph for the Bohr model, however, is to be found in the precision with which it predicts the spectral lines for hydrogen. Allowed energy levels are obtained by substituting values for the principle quantum number n into Eq. (6.14). Thus :

$$E = -R, -R/4, -R/9, -R/16, -R/25, ..., 0.$$

Note that as n gets larger the energy *algebraically* increases. Energies of the electron in allowed orbits are graphically depicted in Fig. 6.6.

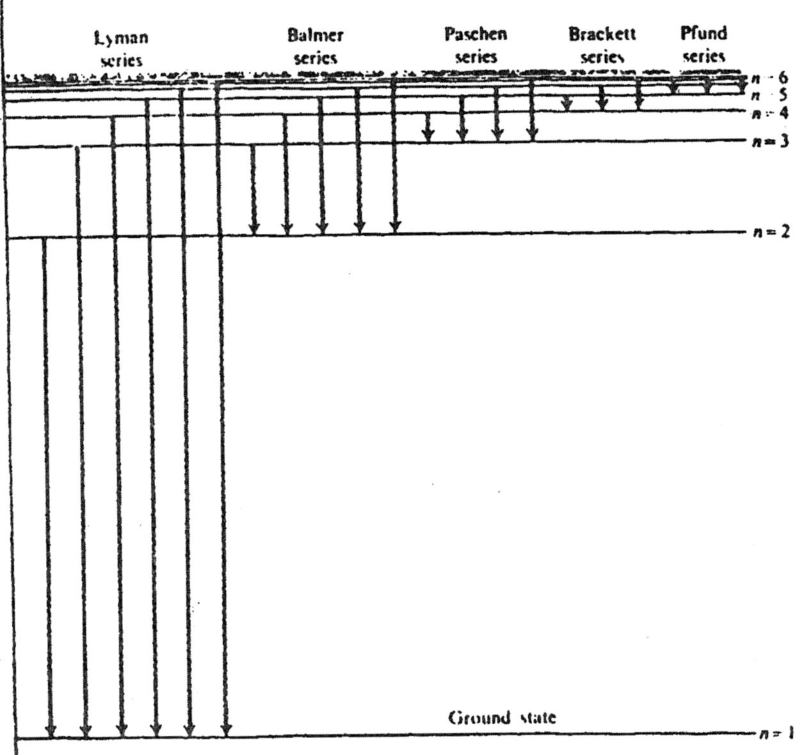

Fig. 6.6. Total energies for the electron in allowed Bohr orbits and energy drops responsible for observed spectral lines.

According to the fourth postulate of Bohr's theory, the energy emitted in the transition of a single electron from an initial stationary state in which the energy is E_{n_2} to a final stationary state in which the energy is E_{n_1} is given by

$$\varepsilon = h\nu = E_{n_2} - E_{n_1}, \qquad (6.18)$$

where ν is the frequency of the corresponding line observed in the spectrum. Substitution of Eq. (6.14) into Eq. (6.18) yields

$$h\nu = -R\left(\frac{1}{n_2^2} - \frac{1}{n_1^2}\right) = R\left(\frac{1}{n_1^2} - \frac{1}{n_2^2}\right)$$

where n_2 and n_1 are the initial and final quantum numbers respectively $n_2 > n_1$. But $\nu = c/\lambda = c\bar{\nu}$, so that

$$\bar{\nu} = \left(\frac{R}{ch}\right)\left(\frac{1}{n_1^2} - \frac{1}{n_2^2}\right) \tag{6.19}$$

or

$$\bar{\nu} = \left(\frac{2\pi^2 m_e e^4}{ch^3}\right)\left(\frac{1}{n_1^2} - \frac{1}{n_2^2}\right) \tag{6.20}$$

The form of Eq. (6.20) is identical to that of Eq. (6.2), the experimentally deduced Rydberg equation, which was

$$\bar{\nu} = R_H\left(\frac{1}{n_2^2} - \frac{1}{n_1^2}\right) \tag{6.2'}$$

A major test of the validity of the Bohr model may thus be made by comparing the calculated value of $2\pi^2 me^4/ch^3$, which is given in Eq. (6.20), with the experimental value for R_H. However, because the Rydberg constant may be measured experimentally with such great precision, it is important that we first make a minor correction in the mathematical treatment in order to account for the finite mass of the nucleus.

Series Limit

Each spectral series has a limit corresponding to the highest value of $\bar{\nu}$, i.e., of the highest energy of transition, when the electron transition occurs between E_{n_1} and E_∞. The wave number of the series limit may be obtained by putting $n_2 = \infty$ in the equation

$$\bar{\nu} = R_H\left(\frac{1}{n_1^2} - \frac{1}{n_2^2}\right)$$

Thus

$$\bar{\nu}_{\text{limit}} = R_H\left(\frac{1}{n_1^2} - \frac{1}{\infty^2}\right) \text{ or } R_H\left(\frac{1}{n_1^2}\right) \text{ since } \frac{1}{\infty^2} = 0$$

$n_1 = 1$ for the Lyman series, 2 for Balmer series and so on.

Beyond the series limit, there exists the continuous spectrum.

Example 6.2. Calculate the wavelength of the first of the Lyman series ($n_1 = 1$ and $n_2 = 2$) in the hydrogen atom emission spectrum.

Solution : Substituting $n_1 = 1$ and $n_2 = 2$ in the expression

$$\bar{\nu} = \frac{1}{\lambda} = R_h = \left(\frac{1}{n_1^2} - \frac{1}{n_2^2}\right)$$

We get

$$\bar{\nu} = (1.097 \times 10^7 m^{-1})\left(\frac{1}{1^2} - \frac{1}{2^2}\right) = 8.228 \times 10^6 m^{-1}$$

Hence,

$$\lambda = \frac{1}{8.228 \times 10^6 m^{-1}} = 1.215 \times 10^{-7} m$$

Example 6.3. Calculate the wavelength, in nanometers, of the first line of the Balmer series, assuming that $n_2 = 3$. Also evaluate the wavelength for the series limit.

Solution : We have

$$\bar{\nu} = R_H\left(\frac{1}{n_1^2} - \frac{1}{n_2^2}\right)$$

For the Balmer series $n_1 = 2$, we can write

$$\frac{1}{\lambda} = (1.097 \times 10^7 m^{-1})\left(\frac{1}{2^2} - \frac{1}{3^2}\right) = 1.524 \times 10^6 m^{-1}$$

Thus,

$$\lambda = \frac{1}{1.524 \times 10^6 m^{-1}} = 6.562 \times 10^{-7} m$$

Converting to namometers, we get

$$\lambda = (6.562 \times 10^{-7} m) = 6.562 \times 10^{-9} m = 6.562 \text{ nm}$$

For evaluating $\bar{\nu}_{\text{limit}}$, We put $n_2 = \infty$. Thus

$$\frac{1}{\lambda_{\text{limit}}} = \bar{\nu}_{\text{limit}} = R_H\left(\frac{1}{n_1^2} - \frac{2}{\infty^2}\right)$$

$$= R_H \times \frac{1}{2^2}$$

$$\frac{1}{\lambda} = \frac{1.097 \times 10^7 m^{-1}}{4} = 2.74 \times 10^6 m^{-1}$$

or $$\lambda = 3.65 \times 10^{-7} m$$

Converting to nanometers, we get

$$\lambda = (3.65 \times 10^{-7} m) = 365 \times 10^{-9} m = 365 \text{ nm}$$

Limitations of Bohr's Theory

Bohr's model was highly successful in accounting for the main features of the hydrogen spectrum as also for the spectrum of other single electron species such as He^+, Li^{2+}, etc., which like hydrogen have a single electron. Yet it also met with certain difficulties. These are :

1. When the hydrogen spectrum was observed with the help of instruments of high resolving powers, it was observed that the individual spectral lines consisted of several fine lines lying close to each other. This is the fine spectrum. Thus the H_a line in the Balmer series of hydrogen spectrum consists of five components closely spaced together, Bohr's theory could not explain the fine structure of hydrogen spectrum.

The fine structure in the line spectrum of hydrogen implies that there are several energy levels close together rather than a single level for each value of n.

2. Another major limitation of Bohr's theory is its incapability n explaining the spectrum of multi-electron atoms. Bohr's theory could not predict the energy states of more complicated atoms (having more than one electron).

3. It is found that, in the presence of a magnetic field, the spectral lines are further split up. This is known as the *Zeeman effect.* Bohr's theory offers no explanation for this.

4. Bohr assumed the electron to be a particle. It is now known that the electron has a dual character. Further, Bohr assumed that an electron revolves in a fixed orbit at fixed distance from the nucleus with a definite velocity.

5. The intensities of spectral lines cannot be calculated.

6. Bohr's theory provides no basis for the classification of elements or for the periodicity in the properties of elements.

Extension of Bohr's Theory

The Bohr Theory was certainly a great advance in the development of the atomic model and was soon followed by modifications considered necessary to explain fine-line structure observed under varying conditions. Sommerfeld, in particular, extended the Bohr model to accommodate elliptical orbits having quantized degrees of eccentricity, which required the introduction of a second quantum number k having the allowed integral values, 1, 2, ..., n.. The shapes of the corresponding orbits are given by

$$\frac{n}{k} = \frac{a}{b},$$

where a is the length of the major axis and b is the length of the minor axis. Thus, when k is equal to n the Bohr-Sommerfeld orbit is a circle. Sommerfeld also showed that if one considers relativity, the energy $E_{n,k}$ for an electron in a given stationary elliptical orbit depends not only on n but also to a very slight degree on k, that is,

$$E_{n,k} = \frac{-2\pi^2 me^4}{n^2h^2}\left[1 + \frac{\alpha^2}{n}\left(\frac{1}{k} - \frac{3}{4n}\right)\right],$$

where $\alpha = 2\pi e^2/hc = 7.2977 \times 10^{-3}$ is called the *Sommerfeld fine structure constant.* The use of Sommerfeld corrected energies rather than Bohr energies in the calculation of spectral wavelengths result in very slight shifts from the values given by Eq. (6.2). In particular, the use of the n different allowed k values associated with a given single value of the principal quantum number leads to a group of very closely spaced lines in the spectrum at those positions for which the uncorrected Bohr theory predicts single lines. In spite of several important modifications of the Bohr model which were to follow in the years between 1913 and 1925, even the revised model was not able successfully to interpret the spectra of atoms containing more than one electron. Nor was it able to account for relative intensities of different spectral lines, or to explain the binding of atoms in molecules. Clearly, some fundamental changes in the model of the atom were in order.

MATTER WAVES

In the mathematical treatment of blackbody radiation, the photoelectric effect, and the Bohr model for the hydrogen atom, we have observed that integers appear as necessary parts of the solutions.

There is another well-known branch of physics in which integers appear naturally in solutions and in which only certain discrete values of physical properties are allowed, that is, classical wave theory. In our discussion of blackbody radiation we have stated that persistent waves in a one-dimensional enclosure are restricted to those for which the distance between the boundary walls is an integral multiple of one-half the wavelength. The audio frequencies emitted by violin strings are not continuous but are limited to certain frequencies referred to as fundamentals and overtones. *Is it possible that the quantum restrictions we have observed are caused by a fundamental wave nature associated with all particles*?

The photon apparently has both particle and wave properties associated with it. At least, in order to explain diffraction we find it convenient to think of electromagnetic radiation as a wave, and in order to explain the photoelectric effect we find it convenient to think of radiation as a stream of discrete particles.

The energy ε of a photon is related to the frequency of the associated electromagnetic wave by the Planck-Einstein relationship

$$\varepsilon = h\nu. \tag{6.21}$$

But we also note that according to the theory of relativity, the relativistic energy of the photon is

$$\varepsilon = mc^2, \tag{6.22}$$

where m is the *relativistic* mass of the photon (the *rest* mass of the photon can be shown to be equal to zero) and c is the speed of light. Combination of Eq. (6.21) and Eq. (6.22) leads to

$$mc^2 = h\nu = hc/\lambda,$$

from which

$$\lambda = h/mc. \tag{6.23}$$

The product mc is the momentum of the photon considered as a *particle,* whereas λ is the wavelength of the photon considered as a *wave*.

In a bold move, Louis de Broglie in 1924 intuitively extended the particle-wave relationship for photons to *all particles* (including electrons) and hypothesized that all particles are guided by associated symbolic waves which he called *pilot waves*. Specially, de Broglie postulated that equations of the forms of Eqs. (6.21) and (6.23) are applicable to *all particles* and that the frequency ν *and the wavelength* λ *of the pilot wave associated with a particle of mass m, velocity v, and total relativistic energy E are given by the equations*

$$\nu = E/h \tag{6.24}$$

and

$$\lambda = h/mv. \tag{6.25}$$

Two years after de Broglie's prediction, C. Davisson and L.H. Germer at the Bell Telephone Laboratory demonstrated diffraction of electrons by a crystal of nickel. This sort of behaviour is possible only for waves, and shows conclusively that electrons do have wave properties. Similar diffraction experiments have been successfully performed with other particles, such as neutrons, protons and hydrogen atoms.

Example 6.4. Calculate the wavelength of a neutron moving with a velocity of 2.8 ms^{-1}.

Solution : Applying the de Broglie equation

$$\lambda = \frac{h}{mv}$$

we have

$$\lambda = \frac{6.626 \times 10^{-34} J\, s}{(1.68 \times 10^{-27} kg)(2.8 ms^{-1}}$$

$$= 1.4 \times 10^{-7} m \text{ (since } 1J = 1 \text{ kg m}^2\text{s}^{-2})$$

Example 6.5. Calculate the wavelength of a vehicle of mass 1.32 $\times$ 10^4 kg moving with a velocity of 100 ms^{-1}.

Solution : Applying the de Broglie equation

$$\lambda = \frac{h}{mv}$$

We have

$$\lambda = \frac{6.26 \times 10^{-34} Js}{(1.32 \times 10^4 kg)(100 ms^{-1})}$$

$$= 5 \times 10^{-40} m$$

Note : *The wavelength of a particle of a large mass is very small. Thus the wave-nature of heavy material object is insignificant.*

Example 6.6. The kinetic energy of a 100 g tennis ball is 344.4 × 10^{-2} J. Calculate its wavelength.

Solution : Kinetic energy is given by

$$\frac{1}{2} mv^2 = 344.4 \times 10^{-2} J$$

Since m = 100 g = –0.1 kg, we get

$$v^2 = \frac{2 \times 344.4 \times 10^{-2} J}{0.1} \text{ kg.} = 68.88\ (ms^{-1})^2$$

$$v = 8.29 ms^{-1}$$

Applying the de Broglie equation

$$\lambda = \frac{h}{mv}$$

we get

$$= \frac{6.626 \times 10^{-34} Js}{(0.1\ kg)\ (8.29\ ms^{-1}}$$

$$= 7.99 \times 10^{-34} m$$

Example 6.7. Calculate the de Broglie wavelength of an electron moving at 1% of the speed of light.

Solution : We have

m = 9.11 × 10^{-31} kg

v = 1% of speed of light = 3.00 × 10^{6} ms^{-1}

Applying de Broglie relation

$$\lambda = \frac{h}{mv}$$

we have

$$\lambda = \frac{6.626 \times 10^{-34} Js}{(9.11 \times 10^{-31} kg)\ (3.00 \times 10^{6} ms^{-1})}$$

$$= 2.43 \times 10^{-10}\, m$$

$$= 243 pm$$

The de Broglie relationship is applicable not only to electrons but also to other material objects in motion. It is clear from the relationship that the wavelength λ decreases as either m or v increases. Quite obviously, very massive bodies in motion such as base balls, humans, or motor vehicles are associated with very small wavelengths.

It is important to point out that the wave character of a particle is significant only in describing the motion of a small microscopic body such as an electron, proton, or neutron because of the fact that they have small wavelengths which are of the same magnitude as the dimensions of such particles. The wavelengths associated with macroscopic (large) bodies are very small and are of little significance.

TABLE 6.2
Wavelengths of Various Particles

Particle	*Mass, m* (kg)	*Velocity, v* (ms^{-1})	*Wavelength, λ* (m)
1-volt electron*	9.1×10^{-31}	5.9×10^{5}	12×10^{-10}
100-volt electron	9.1×10^{-31}	5.9×10^{6}	12×10^{-11}
1000-volt electron	9.1×10^{-31}	5.9×10^{7}	12×10^{-12}
100-volt proton	1.67×10^{-27}	1.38×10^{5}	2.9×10^{-12}
100-volt a-particle	6.6×10^{-27}	6.9×10^{4}	1.5×10^{-12}
H_2 molecule (475 K)	3.3×10^{-27}	2.4×10^{3}	8.2×10^{-11}
.22 rifle Bullet	1.9×10^{-3}	3.2×10^{2}	1.1×10^{-33}
Golf Ball	45×10^{-3}	30	4.9×10^{-34}
Base-ball	140×10^{-3}	25	1.9×10^{-34}

*1-Volt electron means electron accelerated through a potential difference of 1 volt. P.

Distinction Between Matter-Waves and Electromagnetic Waves

Matter waves and electromagnetic waves are distinctly different. Two important distinctions are :

1. Matter waves are not associated with any electric or magnetic fields while electromagnetic waves (radiations) are associated with magnetic and electric fields.

2. All electromagnetic waves travel with the same speed c (the speed of light), 3.00×10^8 ms^{-1}, whereas the de Broglie matter waves may have different velocities.

WAVELENGTHS OF ELECTRONS AND OTHER PARTICLES

Let's now consider an electron in permitted orbit. If the properties of the electron are governed by a wave, the only way to avoid destructive interference is to require that the circumference of the orbit contain an integral number of wavelengths :

$$2\pi r = n\lambda. \tag{6.26}$$

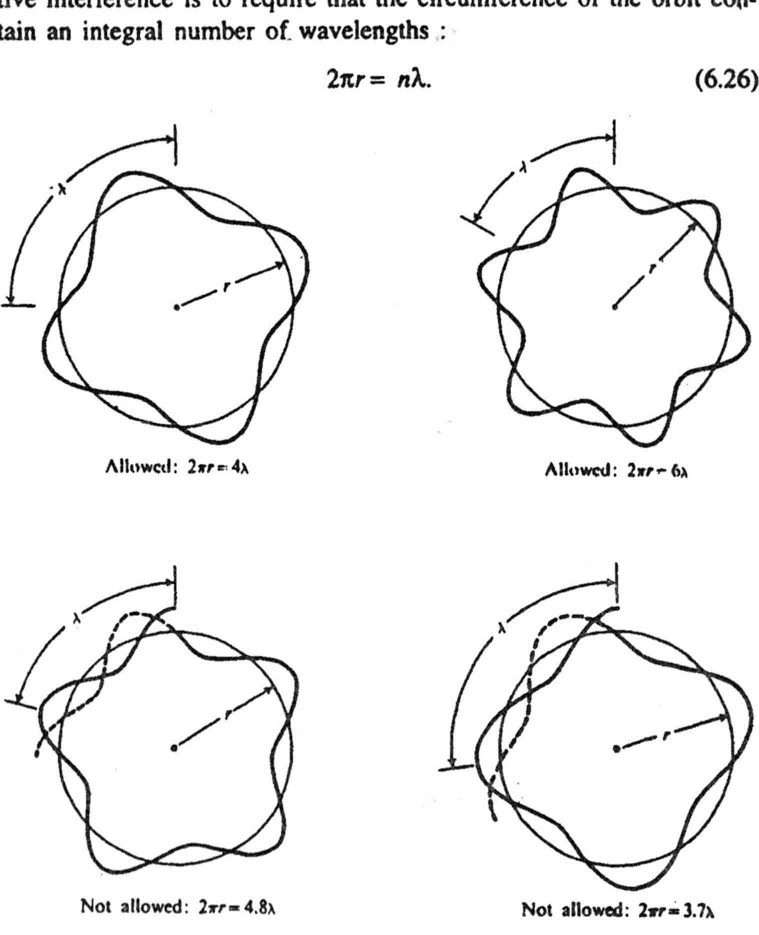

Fig. 6.7. Allowed and nonallowed de Broglie orbits.

Only under such conditions are standing waves produced in a circular path. Wavelengths which do not satisfy this condition will eliminate themselves by destructive interference. Examples of allowed and non-allowed electron waves are shown in Fig. 6.7. The problem of fitting standing electron waves into orbits is the same as that of fitting standing acoustical waves into a metal hoop which is struck with a hammer. In the case of the metal hoop, only those audio frequencies are heard for which standing waves may persist. As in the case of the plucked violin string, persistent vibrations consist of discrete allowed frequencies referred to as *fundamental modes.*

A surprising and important consequence of de Broglie's postulate results when we combine Eqs. to eliminate λ, assuming that the rest mass m_c of the electron may be substituted for the relativistic mass at orbital velocities :

$$m_c vr = nh/2\pi \tag{6.27}$$

Equation (6.27) is *identical to Bohr's third postulate.* By attributing wave properties to the electron, de Broglie was able to justify the bold empirical postulate made by Bohr a decade earlier.

The fact that electrons do indeed have associated wavelengths was experimentally verified in 1927 by Davisson and Germer, who computed the wavelength for electrons from the observed diffraction pattern from a nickel surface. The calculated wavelength for electrons at the experimental velocity exactly agreed with that predicted by the de Broglie relationship, Eq. (6.25).

The de broglie equations are also applicable to particles other than electrons. Knowing the mass and velocity of *any* particle, we may directly calculate its associated wavelength from Eq. (6.25). For example, an average hydrogen molecule at 200°C has a velocity of 2.4 $\times$ 10^5 cm/sec. Since the mass of the hydrogen molecule is 3.3 $\times$ 10^{-24} g, its associated wavelength is calculated to be

$$\lambda = \frac{h}{mv} = \frac{6.62 \times 10^{-27}\ \text{g–cm}^2\text{–sec}^{-1}}{(3.3 \times 10^{-24}\text{g})\,(2.4 \times 10^4\ \text{cm–sec}^{-1})}$$

$$= 0.83 \times 10^{-8}\ \text{cm} = 0.83\ \text{Å}.$$

In like manner, we may calculate that the wavelength of a 2-ton automobile traveling 20 mph is 4.1 $\times$ 10^{-28} Å! This is a ridiculously small wavelength which cannot be measured by any known technique! A few simple calculations of this type immediately disclose that de Broglie wavelengths are experimentally detectable *only* for *very small particles.*

7

Particle Waves and Quantum Mechanics

In our study of classical waves in the chapters 3 and 4 we have been concerned mostly with the expression and evaluation of displacement as a function of position and time. The total displacement function $\Phi(x, y, z, t)$ for a three-dimensional wave must be determined by finding a suitable function which satisfies both the general differential wave equation *and* any boundary conditions imposed in a specific situation.

For all of the classical waves which we have studied, the satisfactory displacement function has in effect been a complete source of all information concerning the wave. For example, the displacement function $y(x, t)$ for the standing wave in the violin string (Eq. 3.51) contains all measurable information concerning the wave. It permits us through proper operation, to calculate displacement at any point in space or time. It also allows the calculation of the frequencies of the infinite number of allowed modes of vibration, from which all possible motions of the string may be synthesized by simple superposition. The satisfactory displacement equation for a wave represents the clearest and most concise definition of the wave which one can construct and represents the essence of all information concerning the wave motion.

DISPLACEMENT FUNCTIONS FOR PARTICLE WAVES

If, as de Broglie postulated, all particles have symbolic pilot waves associated with them whose frequencies are given by $\nu = E/h$ and whose wavelengths are given by $\lambda = h/mv$, then each individual particle must have associated with it some property which is analogous to the classical displacement. In fact, we might expect to find an appropriate displacement-analog function $\Psi(x, y, z, t)$ which completely describes the particle. We might also expect *in cases involving stationary states,* to be able to separate the general function $\Psi(x, y, z, t)$ into a space-dependent amplitude function $\psi(x, y, z)$ and a time- dependent function $f(t)$, so that

$$\Psi(x, y, z, t) = \psi(x, y, z)f(t). \tag{7.1}$$

this type of expectation is the basis for the first postulate of quantum mechanics :

FIRST POSTULATE OF QUANTUM MECHANICS

The state of a single particle in a particular system is described as fully as possible by an appropriate state function, $\Psi(x, y, z, t)$, *which may be expressed in certain particular cases as the product of a time- dependent function and a time-independent amplitude function,* $\psi(x, y, z)$. *Both functions,* Ψ *and* ψ, *must be single- values, continuous, and finite for all values of their coordinates, and must be smoothly varying within the boundaries at which they go to zero.*

A function which is single-valued, continuous, finite, and smoothly varying is said to be *well-behaved.* That is, if it is single-valued and continuous, it has only one value at any given point in coordinate space and there are no gaps or discontinuities in the function. Smooth variation of the function is assured by requiring that its first and second derivatives be continuous so that there will be no sharp angular changes between the boundaries at which the function goes to zero. Some examples of *poorly behaved* functions are shown in Fig. 7.1.

The argument for the first postulate, will have to lie in the success with which resultant equations predict the experimental measurement of physical properties. At this point, we know neither the form nor the significance of the state function Ψ for a particle such as an electron. The classical displacement function $\Phi(x, y, z, t)$ is separable into the product of a space-dependent function, $\psi(x, y, z)$, and a time-dependent function, $e^{-i2\pi\nu t}$, *in those specific cases in which stationary waves exist* and in which the values of ν are *restricted to discrete constant values.* Because of the direct relationship between the total energy of a particle and the frequency of its associated pilot wave, that is, $E = h\nu$, one might expect an analogous relationship to exist for particle waves. Let us propose that the state function $\Psi(x, y, z, t)$ for a particle is separable into the product of a space-dependent function $\psi(x, y, z)$ and a time-dependent function $f(t)$ *in those specific cases in which stationary energy states exist and in which E is restricted to discrete constant values.* This will be justified through later mathematical treatments and results.

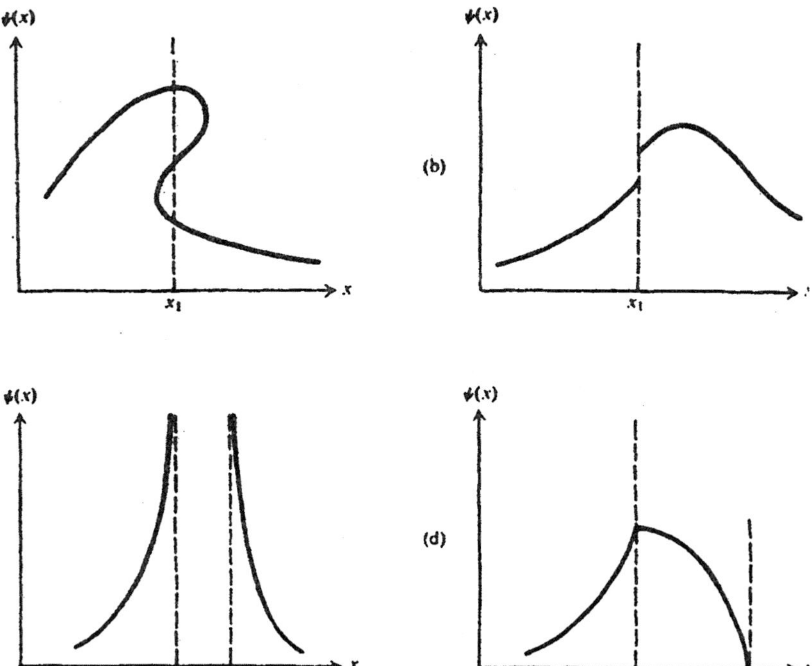

Fig. 7.1. Four examples of poorly behaved functions. (a) Function is not single valued. ψ has three values at x_1. (b) Function is not continuous at x_1. (c) Function is not finite for all values of x, but goes to infinity between x_1 and x_2. (d) Function is not smoothly varying; that is, $d\psi/dx$ is not continuous at x_1. However, the discontinuity in slope at x_2 is allowed, since this is one of the boundaries at which ψ goes to zero.

For the classical stationary wave, the time average value of the square of the absolute displacement, $|\Phi|^2$, (given by $\psi^*\psi$) is a measure of the energy density at a given point in space. *We might also expect that $\psi^*\psi$ will have an equally important meaning for stationary particle waves.* Let then assume a particular case in which $\Psi(x, y, z, t)$

may be expressed as the product of a time-independent function, $\psi(x, y, z)$, multiplied by a time-dependent function $f(t)$, and apply the time-independent classical equation, Eq. (4.14), to the behaviour of the pilot wave associated with a single particle. Substitution of the wavelength of the pilot wave, $\lambda = h/mv$, directly into Eq. (4.14) yields

$$\nabla^2\psi = -4\pi^2 m^2 v^2 \psi / h^2, \tag{7.2}$$

where m is the mass of the particle and v is its velocity. Equation (7.2) is the quantum-mechanical analog of Eq. (4.14), the time-independent classical amplitude equation.

Eq. (7.2) can be rearranged to

$$\nabla^2\psi = \frac{-8\pi^2 m}{h^2}\left(\frac{mv^2}{2}\right)\psi. \tag{7.3}$$

But the kinetic energy T of the particle is $mv^2/2$, so that Eq. (7.3) may be written as

$$\left(-\frac{h^2}{8\pi^2 m}\nabla^2\right)\psi = T\psi. \tag{7.4}$$

Eq. (7.4) may be solved directly only when we specify the required behavioral proprieties and boundary conditions imposed on ψ in a particular problem. When certain conditions are stated, we find that solutions exist only for a set of privileged discrete constant values for T, which are the allowed eigenvalues. The corresponding ψ- functions are the acceptable eigenfunctions of the operator, $(-h^2/8\pi^2 m)\nabla^2$. That is Eq. (7.4) may be solved as an eigenvalue equation whenever T is restricted to a constant value. Furthermore, the set of eigenvalues which we obtain for the kinetic energy are the only allowed values of kinetic energy available to the particle under the specific conditions of the problem. The operator in Eq. (7.4) is called the *kinetic-energy operator* $\hat{T}$ so that

$$\hat{T}\psi = T\psi, \tag{7.5}$$

where

$$\hat{T} = \frac{-h^2}{8\pi^2 m}\nabla^2 \tag{7.6}$$

The actual form of the function ψ will have to determined by trial of likely functions.

In order to develop another operator, Eq. (7.6) may be written as

$$\hat{T} = \frac{1}{2m}\left[-\frac{h^2}{4\pi^2}\left(\frac{\partial^2}{\partial x^2}\right) - \frac{h^2}{4\pi^2}\left(\frac{\partial^2}{\partial y^2}\right) - \frac{h^2}{4\pi^2}\left(\frac{\partial^2}{\partial z^2}\right)\right], \tag{7.7}$$

and the kinetic energy T for the single particle in terms of the component velocities may be expressed as

$$T = \frac{mv^2}{2} = \frac{m}{2}(v_x^2 + v_y^2 + v_z^2), \tag{7.8}$$

or

$$T = \frac{1}{2m}(m^2v_x^2 + m^2v_y^2 + m^2v_z^2). \tag{7.9}$$

But the linear momentum of the particle is given as

$$p = mv,$$

thus,

$$T = \frac{1}{2m}(p_x^2 + p_y^2 + p_z^2), \tag{7.10}$$

where p_x, p_y, and p_z are the component linear momenta in the x-, y-, and z- directions, respectively. Substitution of Eqs. (7.7) and (7.10) into Eq. (7.5) gives

$$\left[-\frac{h^2}{4\pi^2}\left(\frac{\partial^2}{\partial x^2}\right) - \frac{h^2}{4\pi^2}\left(\frac{\partial^2}{\partial y^2}\right) - \frac{h^2}{4\pi^2}\left(\frac{\partial^2}{\partial z^2}\right)\right]\psi = (p_x^2 + p_y^2 + p_z^2)\psi. \tag{7.11}$$

Since x, y, and z are independent variables, each of the operator terms for a given coordinate direction on the left-hand side of Eq. (7.11) corresponds to the linear momentum in that same direction in the expression on the right-hand side. We may also write

$$\left[-\frac{h^2}{4\pi^2}\left(\frac{\partial^2}{\partial x^2}\right)\right]\psi = p_x^2\psi. \tag{7.12}$$

Eq. (7.12) can be solved only after we state the required behavioral and boundary conditions which apply to ψ in a particular problem. Under certain conditions, we find that solutions exist only for a set of privileged eigenvalues for p_x^2. These eigenvalues must be obtained by solving Eq. (7.12) as an eigenvalues equation, this means that we must find acceptable eigenfunctions ψ for the operator, $-h^2/4\pi^2)(\partial^2/\partial x^2)$, under a specific set of problem conditions. If, under the particular problem conditions, p_x^2 is restricted to stationary states, the eigenvalues given by Eq. (7.12) will represent the only stationary-state values of p_x^2 allowed for the particle. The operator $-h^2/4\pi^2)\,\partial^2/\partial x^2$ is given the symbol $\hat{p}_x^2$, Eq. (7.12) can thus be written as :

$$\hat{p}_x^2\psi = p_x^2\psi.$$

If we consider the operator $\hat{p}_x^2$ as consisting of two identical repetitive suboperators such that

$$\hat{p}_x^2 = -\frac{h^2}{4\pi^2}\left(\frac{\partial^2}{\partial x^2}\right) = \frac{h^2}{i^2 4\pi^2}\left(\frac{\partial^2}{\partial x^2}\right) = \left[\frac{h^2}{i2\pi}\left(\frac{\partial}{\partial x}\right)\right]\left[\frac{h^2}{i2\pi}\left(\frac{\partial}{\partial x}\right)\right], \quad (7.14)^*$$

where $i^2 = -1$. Then

$$\hat{p}_x^2 = \hat{p}_x \cdot \hat{p}_x, \tag{7.15}$$

where the operation $\hat{p}_x \cdot \hat{p}_x$ represents two successive suboperations in which

$$\hat{p}_x = \frac{h}{i2\pi}\left(\frac{\partial}{\partial x}\right) \tag{7.16}$$

Similarly, for the y- and z-directions, we may write

$$\hat{p}_y = \frac{h}{i2\pi}\left(\frac{\partial}{\partial y}\right) \tag{7.17}$$

$$\hat{p}_z = \frac{h}{i2\pi}\left(\frac{\partial}{\partial z}\right) \tag{7.18}$$

The operators $\hat{p}_x, \hat{p}_y$, and $\hat{p}_z$ are the *linear-momentum operators* for each of the coordinate directions. For *stationary states* in which the linear momentum in a given direction is restricted to constant values, we may write the eigenvalue equations as follows :

$$\hat{p}_x\psi = p_x\psi, \tag{7.19}$$

$$\hat{p}_y\psi = p_y\psi, \tag{7.20}$$

$$\hat{p}_z\psi = p_z\psi. \tag{7.21}$$

In such stationary-state problems, the operation of the linear-momentum operator on acceptable eigenfunctions yields the allowed values of the linear momentum as eigenvalues.

THE HAMILTONIAN OPERATOR FOR TOTAL ENERGY

The total *nonrelativistic* energy, E, of a single particle is given as the sum of the kinetic energy T and the potential energy V :

$$T + V = E. \tag{7.22}$$

The potential energy V may be a function of x, y, z, and t. However, unless light or other radiation happens to be shining on the system, most problems involving atoms and molecules deal with particles moving in force fields in which V is dependent on *position only* and is independent of time. Such a field in which V depends only on position and is independent of time is called a *conservative force field.*

Multiplication of Eq. (7.22) by ψ gives

$$T\psi + V\psi = E\psi. \tag{7.23}$$

Combination with Eq. (7.5) gives

$$\hat{T}\psi + V\psi = E\psi, \tag{7.24}$$

or

$$(\hat{T} + V)\psi = E\psi, \tag{7.25}$$

which may also be written *in a conservative force field,* where V does not depend on t, as

$$\left[\frac{-h^2}{8\pi^2 m}\nabla^2 + V(x, y, z)\right]\psi = E\psi. \tag{7.26}$$

Since E is constant, Eq. (7.26) is an eigenvalue equation which provides a means for calculating allowed energy eigenvalues for the particle. Equation (7.26) is called the *time- independent Schrodinger equation* or the *Schrodinger amplitude equation.* According to Eq. (7.26) the *time- independent* operator

$$\left[\frac{h^2}{8\pi^2 m}\nabla^2 + V(x, y, z)\right]$$

*The choice of positive roots rather than negative roots for $\hat{p}_x$ in Eq. (7.14) is arbitrary. It is important, however, that we remain consistent in related expressions.

when operating on acceptable eigenfunctions ψ generates, *as eigen-values* in a *conservative* system, the allowed stationary-state energy levels available to the particle. This very special operator is called the *Hamiltonian operator* and is given the symbol $\hat{H}$. Thus

$$\hat{H}\psi = E\psi, \tag{7.27}$$

where

$$\hat{H} = \left[\frac{h^2}{8\pi^2 m} \nabla^2 + V(x, y, z) \right]. \tag{7.28}$$

In order to define $\hat{H}$, we must know the potential energy of the particle as a function of position in the particular conservative force field which applies in a particular problem. Once the form for $V(x, y, z)$ is known, we may write $\hat{H}$ and then try to discover suitable amplitude functions ψ which satisfy both Eq. (7 26) *and* any boundary conditions which we might also impose on the particle.

SOME PROPERTIES OF LINEAR OPERATORS

Before time dependence is introduced into the Schrodinger equation, it is required that a very important property of operators is considered that property is *linearity*. An operator $\hat{A}$ is said to be linear if it obeys the relationship

$$\hat{A}(\Psi_1 + \Psi_2 + \Psi_3 + \ldots) = \hat{A}\Psi_1 + \hat{A}\Psi_2 + \hat{A}\Psi_3 + \ldots \tag{7.29}$$

For example, the operators c (a constant), x, d/dx, and d^2/dx^2 are linear, whereas the operator $\angle$ is nonlinear, since it does not obey Eq. (7.29). Now consider a linear-operator equation of the form

$$\hat{A}\Psi + \hat{B}\Psi + \ldots = 0, \tag{7.30}$$

where $\hat{A}$, $\hat{B}$,... are *different* linear operators. When the constant on the right-hand side of such a linear-operator equation is zero, that is, when Ψ appears to the first degree in each term, the equation is said to be *homogeneous.*

Relationship which Applied to Linear Homogeneous Equations :

If Ψ_1, Ψ_2,... individually are satisfactory solutions for a given linear homogenous equation, then any linear combination of Ψ_1, Ψ_2,... is also a satisfactory solution.

To prove the above relationship, consider two functions, Ψ_1 and Ψ_2, which individually satisfy the same linear homogeneous equation:

$$\hat{A}\Psi_1 + \hat{B}\Psi_1 = 0, \quad (7.31)$$

$$\hat{A}\Psi_2 + \hat{B}\Psi_2 = 0, \quad (7.32)$$

A general form of linear combination of the two functions is

$$a\Psi_1 + b\Psi_2,$$

where a and b are constants.

To show that the linear-combination function, $a\Psi_1 + b\Psi_2$, also satisfies the same linear homogeneous equation consider the Eqs. :

$$\hat{A}(a\Psi_1 + b\Psi_2) + \hat{B}(b\Psi_1 + b\Psi_2) = 0, \quad (7.33)$$

$$a\hat{A}\Psi_1 + b\hat{A}\Psi_2 + a\hat{B}\Psi_1 + b\hat{B}\Psi_2 = 0,$$

$$a(\hat{A}\Psi_1 + \hat{B}\Psi_1) + b(A\Psi_2 + \hat{B}\Psi_2) = 0. \quad (7.34)$$

Substitution of Eqs. (7.31) and (7.32) into Eq. (7.34) yields

$$a(0) + b(0) = 0,$$

which proves the relationship.

DEVELOPMENT OF THE GENERAL TIME-DEPENDENT WAVE EQUATION

First of all, *whatever the form of the general time dependent wave equation, it must reduce to the form of Eq.* (7.26) *in the special case of a conservative system,* in which E is constant. Moreover, in the *general* case, where V is a function of x, y, z, and t, the total energy E is *not* a constant but rather is a function of t. *This means that E itself must not appear as a constant in the general time-dependent equation.* We expect that the general quantum-mechanical wave equation will contain a *time- dependent* operator. Since E is constant in conservative force fields, *it is necessary that the time-dependent operator yield allowed constant values of E as eigenvalues in the special case of a conservative system.*

Consider the following form for the time-dependent differential wave equation for a singe particle of mass m :

$$\left[\frac{-h^2}{8\pi^2 m}\nabla^2 + V(x, y, z, t)\right]\Psi = \hat{E}\Psi, \tag{7.35}$$

where $\hat{E}$ is a *time-dependent energy operator* such that, in *conservative* systems, where E is constant,

$$\hat{E}\Psi + E\Psi. \tag{7.36}$$

Thus, in conservative force fields, where $V(x, y, z, t)$ becomes $V(x, y, z)$, Eq. (7.36) can be substituted into Eq. (7.35) so that

$$\left[\frac{-h^2}{8\pi^2 m}\nabla^2 + V(x, y, z, t)\right]\Psi = E\Psi.$$

Substitution of Eq. (7.1) in the above equation gives

$$\left[\frac{-h^2}{8\pi^2 m}\nabla^2 + V(x, y, z, t)\right][\psi f(t)] = E[\psi f(t)].$$

But since the operator is time independent, $f(t)$ can be factored from each side so that

$$\left[\frac{-h^2}{8\pi^2 m}\nabla^2 + V(x, y, z,)\right]\psi = E\psi$$

or

$$\hat{H}\psi = E\psi,$$

Thus it is proved that the general equation reduces to the time- independent Schrodinger amplitude equation (Eq. 7.26). This means that, we may meet the requirement of reducibility from the general case to the conservative system case if we postulate the relationship given by Eq. (7.36).

Let us now determine the form of the time-dependent energy operator $\hat{E}$. Recall that many classical phenomena, such as the diffraction of light, are explained by assuming that the total wave displacement at any given coordinate position may be considered to be a linear combination of the displacement of several constituent waves (the principle of superposition). By analogy, in order to provide a mathematical scheme capable of describing interference phenomena observed for particle waves the condition is *that $\hat{E}$ be a linear operator,* so that Eq. (7.35) then becomes a linear homogeneous equation,

$$\hat{H}\Psi - \hat{E}\Psi = 0,$$

That is, if Ψ_1 and Ψ_2 individually satisfy the time-dependent differential wave equation (Eq. 7.35), then a linear combination such as $\Psi_1 + \Psi_2$ must also satisfy the *time-dependent* differential wave equation. Linear operators which we might consider are c $c(\partial/\partial t)$, $c(\partial^2/\partial t^2)$, and similar differential operators containing higher- order derivatives, where c is a constant. such operators are linear because

$$c(\Psi_1 + \Psi_2) = c\Psi_1 + c\Psi_2, \tag{7.37}$$

$$c\frac{\partial}{\partial t}(\Psi_1 + \Psi_2) = c\frac{\partial \Psi_1}{\partial t} + c\frac{\partial \Psi_2}{\partial t}, \tag{7.38}$$

$$c\frac{\partial^2}{\partial t^2}(\Psi_1 + \Psi_2) = c\frac{\partial^2 \Psi_1}{\partial t^2} + c\frac{\partial^2 \Psi_2}{\partial t^2}, \tag{7.39}$$

Operators such as $c(\ )^{1/2}$ or $c(\ )^2$ are *not* satisfactory operators, since they are not linear. That is,

$$c(\Psi_1 + \Psi_2)^{1/2} \neq c\Psi_1^{1/2} + c\Psi_2^{1/2}, \tag{7.40}$$

$$c(\Psi_1 + \Psi_2)^2 \neq c\Psi_1^2 + c\Psi_2^2. \tag{7.41}$$

It is then expected that the operator $\hat{E}$ will be of the form c, $c(\partial/\partial t)$, $c(\partial^2/\partial t^2)$, or that it might contain some higher-order time derivative.

The operator c is immediately ruled out because $\hat{E}$ cannot *generally* be a constant since in such a case :

$$c\Psi = E\Psi,$$

which would mean that

$$c = \hat{E} = E,$$

and Eq. (7.35) would under *all* conditions reduce to the conservative system form of Eq. (7.26), which we cannot allow. Thus, $\hat{E}$ *will have to involve a time derivative* and might, for example, be either

$$c\left(\frac{\partial}{\partial t}\right) \text{ or } c\left(\frac{\partial^2}{\partial t^2}\right). \tag{7.42}$$

But how does Ψ depend on t? In order to develop the *time-independent* wave equation, we have had to assume that in a conservative system

$$\Psi(x, y, z, t) = \psi(x, y, z)f(t).$$

Substitution of Eq. (7.1) into Eq. (7.36), which is an eigenvalue equation in conservative system gives

$$\hat{E}\psi(x, y, z)f(t) = E\psi(x, y, z)f(t). \quad (7.43)$$

Since $\hat{E}$ involves a time derivative and not a space derivative, $\psi(x, y, z)$ may be factored from each side, and Eq. (7.43) may then be simplified to

$$\hat{E}f(t) = Ef(t). \quad (7.44)$$

We must now select a suitable form for $f(t)$. For classical waves, time dependence may be expressed conveniently in terms of three basic periodic functions given below :

$$\sin 2\pi\nu t, \text{ or } \cos 2\pi\nu t, \text{ or } e^{\pm 2\pi\nu t}.$$

de broglie's postulates, assign a frequency $\nu = E/h$ and a wavelength $\lambda = h/mv$ to the pilot waves associated with a particle, so that $f(t)$ for periodic dependence in the *pilot wave* will be given by

$$\sin\left(\frac{2\pi Et}{h}\right), \text{ or } \cos\left(\frac{2\pi Et}{h}\right), \text{ or } e^{\pm i2\pi Et/h}, \quad (7.45)$$

or some linear combination of these terms. We may ignore the rest-mass-energy, m_0c^2, and *define E as the classical total energy, T + V.* Of course, such a redefinition of E involves a corresponding change in the calculated value of ν (the frequency of the hypothetical associated pilot wave). This is not required in the Schrodinger theory since ν is not experimentally observable. However, we are able to measure *wavelengths* for particle waves through diffraction measurements.

Let's now substitute the two forms for $\hat{E}$ given by Eq. (7.42) and the three possible basic forms of $f(t)$ given by Eq. (7.45) into Eq. (7.44) in order to determine appropriate forms for both $\hat{E}$ in general and $f(t)$ in conservative systems.

Trial of $c(\partial^2/\partial t^2)$ as the operator with the sine form of $f(t)$:

$$c\frac{\partial^2}{\partial t^2}\left[\sin\left(\frac{2\pi Et}{h}\right)\right] = E\sin\left(\frac{2\pi Et}{h}\right)$$

Double differentiation gives

$$-c\left(\frac{4\pi^2E^2}{h^2}\right)\sin\left(\frac{2\pi Et}{h}\right)= E\sin\left(\frac{2\pi Et}{h}\right),$$

or

$$c + \frac{-h^2}{4\pi^2E}\text{ (tentative)},$$

which would mean that $\hat{E}$ would have to be :

$$\hat{E} = c\left(\frac{\partial^2}{\partial t^2}\right) = \frac{-h^2}{4\pi^2E}\left(\frac{\partial^2}{\partial t^2}\right)\text{ (tentative)}.$$

But such a form for $\hat{E}$ *cannot be allowed since it introduces E as a constant into the general time-dependent form of the wave equation* (Eq. 7.35). E as such must not appear in the correct form for the time-dependent wave equation; E *may appear only in special solutions*!

Trial of $c(\partial^2/\partial t^2)$ as the operator with either cos $(2\pi Et/h)$ or with $e^{+-i2\pi Et/h}$ also leads to a result in which c contains the term E. Furthermore, *all higher-order time derivatives are also unsatisfactory*, because they all introduce E to some *higher* power into the constant c and thus into the general wave equation. Since any other acceptable periodic time function would have to be some linear combination of the three forms given in (7.45), it is clear that $\hat{E}$ cannot be of the form $c(\partial^2/\partial t^2)$, nor can it include a time derivative higher in order than the first derivative.

Trial of $c(\partial/\partial t)$ operator. Neither sin $(2\pi Et^\wedge/h)$ nor cos $(2\pi Et/h)$ is an eigenfunction of $c(\partial/\partial t)$, so that the sine and cosine functions are eliminated, which leaves *only* the function $e^{+-i2\pi Et/h}$. Let's then substitute the *one remaining satisfactory form of* $\hat{E}$, that is $c(\partial/\partial t)$, along with the *one remaining satisfactory form for* $f(t)$, that is $e^{+-i2\pi Et/h}$, into Eq. (7.44) in order to evaluate c :

$$c\left(\frac{\partial e^{\pm i2\pi Et/h}}{\partial t}\right) = Ee^{\pm i2\pi Et/h},$$

$$c\left(\pm\frac{i2\pi E}{h}\right)e^{\pm i2\pi Et/h} = Ee^{\pm i2\pi Et/h},$$

or

$$c = \pm \frac{h}{i2\pi},$$

and therefore

$$\hat{E} = c\left(\frac{\partial}{\partial t}\right) = \pm \frac{h}{i2\pi}\left(\frac{\partial}{\partial t}\right) \tag{7.46}$$

Either the positive or the negative version (Eq. (7.46) *is a satisfactory time-dependent energy operator since neither contains E as a constant.* We'll *arbitrarily* select the negative solution which corresponds to the negative exponential form for $f(t)$ in conservative systems; that is,

$$f(t) = e^{-i2\pi Et/h}, \tag{7.47}$$

so that we may now write, by substitution of Eq. (7.47) into Eq. (7.1),

$$\Psi(x, y, z, t) = \psi(x, y, z)\, e^{-i2\pi Et/h} \tag{7.48}$$

as the *time-dependent wave function* in a conservative system.

Substitution of $\hat{E} = -(h/i2\pi)(\partial/\partial t)$ into Eq. (7.35) yields

$$\left[\frac{-h^2}{8\pi^2 m}\nabla^2 + V(x, y, z, t)\right]\Psi = \frac{-h}{i2\pi}\left(\frac{\partial \Psi}{\partial t}\right) \tag{7.49}$$

which is the *general time-dependent Schrodinger equation.** Note that the complex conjugate of Eq. (7.49) is

$$\left[\frac{-h^2}{8\pi^2 m}\nabla^2 + V(x, y, z, t)\right]\Psi^* = \frac{-h}{i2\pi}\left(\frac{\partial \Psi^*}{\partial t}\right) \tag{7.50}$$

which corresponds to the choice of the positive for $\hat{E}$ in Eq. (7.46). The time-dependent Schrodinger equation is often written in shorter operator form as

$$\hat{H}\Psi = \hat{E}\Psi. \tag{7.51}$$

It is important to note that the general time-dependent *quantum-mechanical* differential wave equation (Eq. 7.49) differs from the *classical* time-dependent differential wave equation in two ways :

1. The time derivative in the quantum-mechanical wave equation is a *first derivative,* whereas the time derivative in the classical wave equation is a second derivative.

---

*E. Schrodinger, *Ann. Physik*, **81**, 109 (1926)

2. The quantum-mechanical wave function (state function) $\Psi(x, y, z, t)$ ***must he complex,*** whereas the classical wave function $\Phi(x, y, z, t)$ may always be real.

The above two important differences between the quantum-mechanical and the classical wave equations result since we adhere in general to de Broglie's two basic postulates, $\nu = E/h$ and $\lambda = h/mv$, with the exception that we redefine E as the total classical energy rather than the total relativistic energy, and in addition that we require all quantum-mechanical operators to be linear.

On the basis of the expected validities of Eqs. (7.49) and (7.26), we can now write the second postulate of quantum mechanics :

SECOND POSTULATE OF QUANTUM MECHANICS

a) *The possible state functions $\Psi(x, y, z, t)$ for a single particle are given by the solution of the time-dependent Schrodinger wave equation* :

$$\left[\frac{-h^2}{8\pi^2 m}\nabla^2 + V(x, y, z, t)\right]\Psi = \left[\frac{-h}{i2\pi}\left(\frac{de}{\partial t}\right)\right]\Psi, \qquad (7.49')$$

or, more generally,

$$\hat{H}\Psi = \hat{E}\Psi. \qquad (7.51')$$

b) *In the special case of conservative systems, where V is not dependent on t, the possible time-independent amplitude functions $\psi(x, y, z)$ for a single particle are given by solution of the time-independent Schrodinger equation,*

$$\left[\frac{-h^2}{8\pi^2 m}\nabla^2 + V(x, y, z, t)\right]\psi = E\psi,$$

or more generally,

$$\hat{H}\psi = E\psi.$$

QUANTUM-MECHANICAL OPERATORS

Let us now show, for a single particle in a *conservative* force field, that the operation of time-independent operators on the *general* function $\Psi(x, y, z, t)$ yields the same eigenvalues as those given by the solutions of the three equations.

$$\hat{T}\Psi = T\Psi \tag{7.5'}$$

$$\hat{p}_x\Psi = p_x\Psi \tag{7.19'}$$

and

$$\hat{H}\Psi = E\Psi \tag{7.27'}$$

For example, substitute Eq. (7.48) into Eq. (7.19′) :

$$\hat{p}_x\left[\frac{\Psi(x, y, z, t)}{\exp(-i2\pi ET/h)}\right] = p_x\left[\frac{\Psi(x, y, z, t)}{\exp(-i2\pi ET/h)}\right].$$

Cancellation of the exponential terms yields

$$\hat{p}_x\Psi = p_x\Psi. \tag{7.52}$$

Furthermore, recall that it is necessary that time-dependent operators, such as $\hat{E}$, operate on $\Psi(x, y, z, t)$ rather than on $\psi(x, y, z)$. For example,

$$\hat{E}\Psi = E\Psi. \tag{7.36'}$$

TABLE 7.1
Some Important Quantum Mechanical Operators

Dynamical variable	Operator symbol	Operation
Position	$\hat{x}$ $\hat{y}$ $\hat{z}$	x (multiplication by x) y (multiplication by y) z (multiplication by z)
Time	$\hat{t}$	t (multiplication by t)
Linear momentum		
x-direction	$\hat{p}_x$	$\frac{h}{i2\pi}\left(\frac{\partial}{\partial x}\right)$
y-direction	$\hat{p}_y$	$\frac{h}{i2\pi}\left(\frac{\partial}{\partial y}\right)$
z-direction	$\hat{p}_z$	$\frac{h}{i2\pi}\left(\frac{\partial}{\partial z}\right)$
Kinetic energy T	$\hat{T}$	$\frac{-h^2}{8\pi^2 m}\nabla^2$
Potential energy V	$\hat{V}$	$V(x, y, z, t)$ (general)

		$V(x, y, z)$ (conservative system) (multiplication by V)
Total energy $T + V$ (Hamiltonian)	$\hat{H} = \hat{T} + \hat{V}$ $= \hat{T} + V$	$\frac{-h^2}{8\pi^2 m}\nabla^2 + V(x, y, z, t)$ (general)
		$\frac{-h^2}{8\pi^2 m}\nabla^2 + V(x, y, z)$ (conservative system)
Total energy $T + V$ (time operator)	$\hat{E}$	$\frac{-h}{i2\pi}\left(\frac{\partial}{\partial t}\right)$

THIRD POSTULATE OF QUANTUM MECHANICS

a) *For every dynamical variable A there is a corresponding linear operator $\hat{A}$. If the dynamical variable is capable of exact experimental determination, its only possible exact values are those given by the eigenvalues of the equation*

$$\hat{A}\Psi = A\Psi. \tag{7.53}$$

b) *In the special case of conservative systems and for time- independent operators, the only possible exact values of the dynamical variable are given by*

$$\hat{A}\psi = A\psi. \tag{7.54}$$

Many of the important quantum-mechanical operators are time- independent and force fields of interest in atomic and molecular systems are often conservative, so that the time-independent relationship given by Eq. (7.54) is very important.

THE SIGNIFICANCE OF ψ IN CONSERVATIVE SYSTEMS

The Schrodinger amplitude equation (E. 7.27), which is an eigenvalue equation in conservative force fields, may be written as

$$\hat{T}\psi + V\psi = E\psi \tag{7.55}$$

Equation (7.55), which may involve complex numbers (ψ may be complex) may be converted to a form involving only real numbers by multiplication by ψ^*. Thus

$$\psi^*\hat{T}\psi + \psi^*V\psi + \psi^*E\psi, \tag{7.56}$$

or

$$\psi^*\hat{T}\psi + V\psi^*\psi = E\psi^*\psi, \tag{7.57}$$

Since $\psi^*\psi = |\psi|^2$, a real number,

$$\psi^*\hat{T}\psi + V|\psi|^2 = E|\psi|^2. \tag{7.58}$$

Since the second and third terms in Eq. (7.58) are real, the first term, $\psi^*\hat{T}\psi$, must also be real. Thus, Eq. (7.58) contains only real quantities.

In order to investigate the real quantity $\psi^*\hat{T}\psi$ more fully, let us consider a method for expressing average quantities.

Let's assume that we wish to calculate the average grade on an examination which has been given to a class of 10 students. Let p_i be the normal probability that a randomly selected student has received the grade G_i. Then we may express the average grade $\overline{G}$ as

$$\overline{G} = \sum_{\substack{\text{all possible}\\ \text{grades}}} G_i p_i. \tag{7.59}$$

In order to understand Eq. (7.59) more clearly, let's consider the following specific grade distribution :

Grade, G_i	Number of students receiving grade	Normal probability, p_i
90	1	0.1
80	2	0.2
70	5	0.5
60	2	0.2
	10	1.0

We'll define the *normal probability* that a randomly selected individual student has received a particular grade as the *fraction of the total* number of students who have received that grade. Such a designation of probability, *although not mathematically required*, ensures that the total probability is unity. That is, the probability that an individual student receives any grade, regardless of what it might be, is unity or certainty. When the sum of all the individual probabilities is defined as unity, the distribution is said to be *normalized*, and the probabilities are *normal probabilities*. Thus, since two students out of 10 have received a grade of 80, the normal probability that a randomly selected student has received a grade of 80 is 0.2. The average grade $\bar{G}$ is then calculated as

$$\bar{G} = \sum G_i p_i = 90(0.1) + 80(0.2) + 70(0.5) + 60(0.2)$$

$$= 9 + 16 + 35 + 12 = 72.$$

In order to extend Eq. (7.59) to an infinite number of possible grades, each involving a differential probability, we can write

$$\bar{G} = \int_{\substack{\text{all possible} \\ \text{grades}}} G\, dp. \tag{7.60}$$

Eq. (7.60) is applicable to the measurement of other average quantities.

The average potential energy $\bar{V}$ for a single particle in three dimensional in a conservative force field, considering all space available to the particle, is given by

$$\bar{V} = \int_{\text{all space}} V(x, y, z)\, dp, \tag{7.61}$$

where dp is the differential probability that the particle has a potential energy between V and $V + dV$. This is the same as the probability that the particle is within the volume element $dx \cdot dy \cdot dz$ at the position x, y, z, since the potential energy is directly related to the position. We may express dp in terms of the probability P that the particle is found in a *unit* volume as follows :

probability that the particle is found in the *differential* volume, $dx \cdot dy \cdot dz$, at position x, y, z	=	Probability per *unit* volume that the particle is found at position x, y, z	×	Volume of the differential element $dx \cdot dy \cdot dz$

$$dp = P(dx \cdot dy \cdot dz). \tag{7.62}$$

We may designate the volume element $(dx \cdot dy \cdot dz)$ as $d\tau$, so that

$$dp = P\, d\tau, \tag{7.63}$$

where P is the probability per unit volume of finding the particle and is called the *probability density*. Substitution of Eq. (7.63) in Eq. (7.61) yields

$$\bar{V} = \int VP\, d\tau = \infty_{-\infty}^{+\infty}\ VP\, d\tau. \qquad (7.64)$$

We may now multiply both sides of Eq. (7.58) by the differential space element $d\tau$,

$$\psi^* \hat{T}\psi\, d\tau + V\, |\psi|^2\, d\tau = E\, |\psi|^2\, d\tau, \qquad (7.65)$$

and integrate over all space available to the particle, recalling that E is constant in a conservative force field,

$$\int_{-\infty}^{+\infty} \psi^* \hat{T}\psi\, d\tau + \int_{-\infty}^{+\infty} V\, |\psi|^2\, d\tau = E \infty_{-\infty}^{+\infty}\ |\psi|^2\, d\tau. \qquad (7.66)$$

A comparison of Eq. (7.64) with the second term on the left-hand side of Eq. (7.66) reveals that $\infty_{-\infty}^{+\infty}\ V\, |\psi|^2\, d\tau$ would be the average potential energy $\bar{V}$ *if* $|\psi|^2$ *were equal to P, the probability density.*

FOURTH POSTULATE OF QUANTUM MECHANICS

a) *The term* $|\Psi|^2\, d\tau$ or $\Psi^*\Psi\, d\tau$ *is the time- independent probability that a single particle exists at a given time t in the space element* $d\tau$, *that is, between,* x, y, z *and* $(x + dx)$, $(y + dy)$, *and* $(z + dz)$.

b) *For the special case of a conservative system, in which the single particle is restricted to an energy eigenstate, the time-independent probability that the particle exists in the space element* $d\tau$ *is* $|\psi|^2\, d\tau$ *or* $\psi^*\psi\, d\tau$.

We may show the relationship between (a) and (b) in the fourth Postulate by noting that $\Psi = \psi \exp(-i2\pi Et/h)$ in conservative force fields. Then, if the energy of the particle is restricted to the single value E,

$$\Psi^*\Psi = \psi^* \exp(i2\pi Et/h)\psi \exp(-i2Et/h) = \psi^*\psi. \qquad (7.67)$$

In a conservative force field that the probability per unit volume at position x, y, z, or *probability density*, may be expressed as the probability that the particle exists in the differential element at position x, y, z divided by the volume of the element or

$$P = \frac{|\psi|^2\, d\tau}{d\tau} = |\psi|^2 = \psi^*\psi. \tag{7.68}$$

Because of its immediate association with the probability of finding a particle, ψ is often called the ***probability amplitude.*** The fourth postulate of quantum mechanics seem intuitively acceptable, since we have already shown in classical mechanics that $|\psi|^2$ represents any of several quantities comparable to the probability of finding a particle. That is, for elastic waves on springs and in strings, $|\psi|^2$ was a direct measure of the energy density at a given point in space. In addition, the square of either the magnetic amplitude or the electrical amplitude at any point in an electromagnetic wave is directly proportional to the *energy density* at that point, and according to the theory of relativity, $m = E/c^2$, so that for particle waves we might expect that $|\psi|^2$ would be a direct measure of *mass density.* That is, in an electromagnetic wave, the energy density may be thought of as a measure of concentration of photons at a given position or as a measure of the probability per unit volume of finding an individual photon at that same position. In the same manner, for an electron wave in a conservative system, the energy density (as given by $|\psi|^2$) is considered a measure of the probability per unit volume of finding the electron at a given position.

Based on Postulate IV, it now follows that, in conservative force fields,

$$\int_{-\infty}^{+\infty} V\,|\psi|^2\, d\tau = \int_{-\infty}^{+\infty} VP\, d\tau = \bar{V}. \tag{7.69}$$

NORMALIZED WAVE FUNCTIONS

Since the single particle under consideration must be found somewhere in space, it is necessary that the total probability of finding the particle in all space, $\int_{-\infty}^{+\infty} |\psi|^2\, d\tau$, must be a finite constant. It is mathematically convenient to define the total probability of finding the particle in all space as unity. According to this definition, when the probability of an event is unity, the event is certain to occur. When the probability of finding a particle in all of space is unity, we are certain to find the particle somewhere in space. We thus desire that the amplitude function, ψ, obey the condition

$$\int_{-\infty}^{+\infty} |\psi|^2 \, d\tau = \text{total probability of finding the particle} = 1, \qquad (7.70)$$

which is called the *normalization condition.* A wave function which obeys Eq. (7.70) is said to be *normalized.*

AVERAGE VALUES FOR DYNAMICAL VARIABLES

Substitution of Eqs. (7.69) and (7.70) into Eq. (7.68) yields, for *normalized amplitude functions,*

$$\int_{-\infty}^{+\infty} \psi^* \hat{T} \psi \, d\tau + \bar{V} = E. \qquad (7.71)$$

But in a conservative force field the total energy E is constant and

$$\bar{T} + \bar{V} = E, \qquad (7.72)$$

where $\bar{T}$ is the average or mean kinetic energy of the particle taken over all *positions in space available to the particle.* In a system of many particles, $\bar{T}$ is the average kinetic energy for all of the particles assuming that all of the space is available to each of the particles. Comparison of Eq. (7.71) and Eq. (7.72) yields, for normalized amplitude functions in a conservative system,

$$\bar{T} = \int_{-\infty}^{+\infty} \psi^* \hat{T} \psi \, d\tau. \qquad (7.73)$$

Note that it is *not* a requirement of Eq. (7.73) that ψ or ψ^* be an eigenfunction of $\hat{T}$

We may now also rewrite Eq. (7.69) for comparison :

$$\bar{V} = \int_{-\infty}^{+\infty} V|\psi|^2 \, d\tau = \int_{-\infty}^{+\infty} \psi^* \hat{V} \psi \, d\tau. \qquad (7.74)$$

When the amplitude functions are *not normalized,* it is easy to show that $\bar{T}$ and $\bar{V}$, in conservative force fields, are given as

$$\bar{T} = \frac{\int_{-\infty}^{+\infty} \psi^* \hat{T} \psi \, d\tau}{\int_{-\infty}^{+\infty} \psi^* \psi \, d\tau}, \qquad (7.75)$$

and

$$\bar{V} = \frac{\int_{-\infty}^{+\infty} \psi^* \hat{V} \psi \, d\tau}{\int_{-\infty}^{+\infty} \psi^* \psi \, d\tau}. \tag{7.76}$$

The similarity of Eqs. (7.75) and (7.76) now leads to another general postulate. Before stating the postulate, however, we note that the operators $\hat{T}$ and $\hat{V}$ in Eqs. (7.75) and (7.76) are time-independent ($\hat{V}$ is time-independent by virtue of the restriction to conservative force fields). It has already been shown that time-*dependent* operators must operate on $\Psi(x, y, z, t)$, the more general state function, rather than on $\psi(x, y, z)$, the time-independent function. Recognizing the greater generality of Ψ over ψ, we then write the fifth postulate of quantum mechanics as follows.

FIFTH POSTULATE OF QUANTUM MECHANICS

a) *The expected mean value $\bar{A}$ of a series of measurements of an observable A made over a large number of particles each in the state represented by Ψ is*

$$\bar{A} = \frac{\int_{-\infty}^{+\infty} \Psi^* \hat{A} \Psi \, d\tau}{\int_{-\infty}^{+\infty} \Psi^* \Psi \, d\tau}, \tag{7.77}$$

where $\hat{A}$ is the quantum mechanical operator for the observable.

b) *For the special case of conservative systems in which the operator $\hat{A}$ does not depend explicitly on time,*

$$\bar{A} = \frac{\int_{-\infty}^{+\infty} \psi^* \hat{A} \psi \, d\tau}{\int_{-\infty}^{+\infty} \psi^* \psi \, d\tau}. \tag{7.78}$$

c) *For the special case of conservative systems and time- independent operators and where the wave functions are normalized.*

$$\bar{A} = \int_{-\infty}^{+\infty} \psi^* \hat{A} \psi \, d\tau. \tag{7.79}$$

Since $\bar{A}$ *represents the probable result of an experimental measurement of* A, it is also called the *expectation value.* The fifth Postulate, called the *mean-value postulate,* allows the calculation of *average* or *mean* values of observables for which stationary-state values cannot be measured exactly. Note that if ψ *happens to be* an eigenfunction of $\hat{A}$, then $\hat{A}\psi = A\psi$, and Eq. (7.78) reduces to

$$\bar{A} = \frac{\int_{-\infty}^{+\infty} \psi^* A\psi \, d\tau}{\int_{-\infty}^{+\infty} \psi^*\psi \, d\tau} = A.$$

That is, *every* measurement of the observable for a particle in the eigenstate represented by ψ is the eigenvalue A for that state, so that the average value is also the eigenvalue.

8

Wave Mechanics of Some Simple Systems

FREE PARTICLE IN ONE DIMENSION

Consider a single particle of mass m which, in the absence of a force field, is free to move anywhere in the x- direction. There are no external forces operating on the particle and the potential energy is a constant which we'll assume to be *zero.* The Schrodinger amplitude equation in one dimension is :

$$\frac{-h^2}{8\pi^2 m}\left(\frac{d^2\psi}{dx^2}\right) = (E - V)\,\psi = E\psi, \tag{8.1}$$

where $E = T + V = T$. Thus the total energy E is given by the kinetic energy T, and is constant in the conservative force field. Equation (8.1) may be rearranged to

$$\frac{d^2\psi}{dx^2} = \frac{-8\pi^2 mE}{h^2}\,\psi, \tag{8.2}$$

which in an eigenvalue equation. The amplitude function ψ, which represents the *complete description of the particle,* must be found by solving Eq. (8.2). It can be shown readily by substitution that either of the following two functions is a satisfactory solution :

$$\psi_1 = A \exp\left[\frac{i2\pi}{h}(2mE)^{1/2}x\right], \tag{8.3}$$

or

$$\psi_2 = B \exp\left[\frac{i2\pi}{h}(2mE)^{1/2}x\right], \tag{8.4}$$

where A and B are constants.

We will now calculate the relative probability density $\psi^*\psi$ (ψ is not normalized) as a function of x, using ψ_1 from Eq. (8.3)

$$\psi_1^*\psi_1 = A^* \exp\left[-\frac{i2\pi}{h}(2mE)^{1/2}x\right] A \exp\left[\frac{i2\pi}{h}(2mE)^{1/2}x\right]$$

$$= A * A = |A|^2 = \text{constant} \tag{8.5}$$

Note that the probability density is independent of x; that is, there is an equal probability of finding the particle at any point along its path from $x = -\infty$ to $x = +\infty$. The particle is totally *nonlocalized.*

Similarly, the use of ψ_2 instead of ψ_1 for the calculation of probability density gives

$$\psi_2^*\psi_2 = B^*B = |B|^2 = \text{constant}.$$

Apparently then, *if* ψ_1 and ψ_2 are both *equally representative* solutions, that is, if the particle has an equal probability of moving in either direction

$$|A|^2 = |B|^2 \tag{8.6}$$

Let's now use the operator postulate (Postulate 2b)

$$\hat{p}_x\psi = p_x\psi$$

in order to determine the allowed exact values for the momentum. The allowed eigenvalues of the observable p_x, are obtained by the operation of $\hat{p}_x$ on either ψ_1 or ψ_2. Operation first on ψ^1 yields

$$\hat{p}_x\psi_1 = \frac{h}{i2\pi}\frac{\partial}{\partial x} A \exp\left[\frac{i2\pi}{h}(2mE)^{1/2}x\right]$$

$$= \frac{h}{i2\pi}\frac{i2\pi}{h}(2mE)^{1/2} A \exp\left[\frac{i2\pi}{h}(2mE)^{1/2}x\right]$$

Cancellation of terms and reinsertion of ψ_1 gives

$$\hat{p}_x\psi_1 = +(2mE)^{1/2}\psi_1,$$

from which

$$\hat{p}_x = +(2mE)^{1/2}. \tag{8.7}$$

Similarly, operation of $\hat{p}_x$ on ψ_2 yields

$$\hat{p}_x\psi_2 = -(2mE)^{1/2}\psi_2,$$

from which

$$p_x = -(2mE)^{1/2}. \tag{8.8}$$

Apparently then, if ψ_1 is chosen as the wave function for the free particle, the momentum, which is a vector quantity, must be *positive* and, since no boundary conditions have been imposed, the momentum may correspond to any exact value whatsoever for the total energy E. (Note that for the classical particle for which $V = 0$).

$$E = T = \frac{mv_x}{2} = \frac{m^2v_x^2}{2m} = \frac{p_x^2}{2m}, \tag{8.9}$$

so that

$$p_x = \pm(2mE)^{1/2}, \tag{8.10}$$

where $(2mE)^{1/2}$ is the *magnitude* of the linear momentum in the x direction. If ψ_2 is chosen to represent the free particle, it is clear from Eq. (8.8) that the momentum must be *negative* and may once again correspond to any exact value whatsoever for the total energy E. Thus, regardless of whether we choose ψ_1 or ψ_2, the energy of the free particle is not quantized. Correspondingly, neither is the momentum as given by Eq. (8.10) quantized. However, if ψ_1 is chosen, the continuous range of exact momentum values available to the particle is restricted to *positive* values. This means that motion of the particle is restricted to the $+x$ direction. On the other hand, if ψ_2 is chosen to represent the free particle, the particle is allowed any exact *negative* value whatsoever for the momentum; the motion of the particle is thus restricted to the $-x$ direction.

A more general wave function which allows the free particle to move in *either* direction may be written as the linear combination of ψ_1 and ψ_2. Thus

$$\psi = \psi_1 + \psi_2. \tag{8.11}$$

ψ is also a solution of the time-independent Schrodinger equation, since ψ_1 and ψ_2, each of which has the *same eigenvalue energy E*, obey the same linear homogeneous equation,

$$\hat{H}\psi - E\psi = 0$$

and therefore any linear combination of ψ_1 and ψ_2 is also a solution. That is,

$$\hat{H}\psi_1 - E\psi_1 = 0, \; \hat{H}\psi_2 - E\psi_2 = 0;$$

therefore,

$$\hat{H}(\psi_1 + \psi_2) - E(\psi_1 + \psi_2) = 0.$$

Wave functions representing states having the same energy are said to be *degenerate.* However, it is important, to note that in the *general* case, if the eigenvalue energies E_1 and E_2 corresponding to the eigenfunctions ψ_1 and ψ_2 are *not equal,* i.e., if the eigenfunctions are *nondegenerate,* each of the functions obeys a *different* linear homogeneous equation :

$$\hat{H}\psi_1 - E_1\psi_1 = 0, \; \hat{H}\psi_2 - E_2\psi_2 = 0,$$

in which case $(\psi_1 + \psi_2)$ is *not* an eigenfunction of $\hat{H}$.

Substitution of Eqs. (8.3) and (8.4) for ψ_1 and ψ_2 in Eq. (8.10) yields

$$\psi = A \exp\left[\frac{i2\pi}{h}(2mE)^{1/2}x\right] + B \exp\left[\frac{-i2\pi}{h}(2mE)^{1/2}x\right]. \quad (8.12)$$

If the particle has an equal probability of moving in either direction, $|A|^2$ must be equal to $|B|^2$ according to Eq. (8.6). On the other hand, if the particle is restricted to movement in the $(+x)$-direction *only,* B must equal zero, in which case the *second* term in Eq. (8.12) disappears. Conversely, if the particle is restricted to motion in the $(-x)$-direction *only,* A must equal zero, in which case the *first* term of Eq. (8.12) disappears.

It must be clearly understood that the energy of a free particle is not restricted to discrete values. The ionized electron is a good example of a free particle in three-dimensional space. Once the electron escapes from the coulombic force field created by the nucleus (once it ionizes) its position is in no way localized and it may assume any exact energy level.

A PARTICLE IN A ONE-DIMENSIONAL BOX

Let us now consider that the particle can move in only the x-direction within a one-dimensional potential energy well of infinite depth, which is referred to as a "box." In order to ensure that the particle remains in the box let us assume that $V = 0$ everywhere in the box and $V = \infty$ everywhere outside the box. Such a box, of length a, (Fig. 8.1). The Schrodinger amplitude equation for this single-particle system is

$$\frac{-h^2}{8\pi^2 m}\left(\frac{d^2\psi}{dx^2}\right) = E\psi, \tag{8.13}$$

where $E = T + V = T$. Equation (8.13) is identical to Eq. (8.1) for the free particle, so that Eqs. (8.3) and (8.4) are valid solutions.

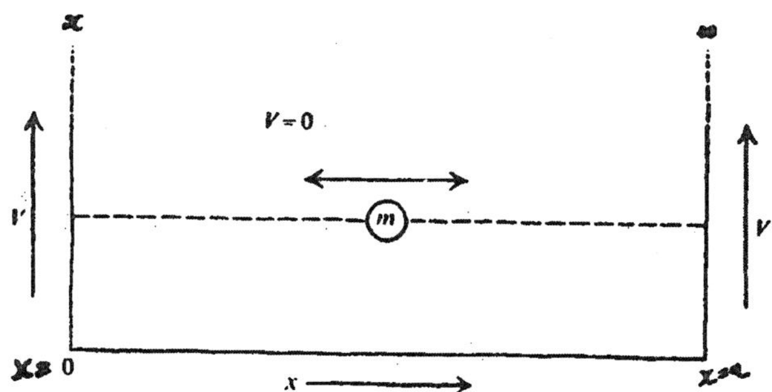

Fig. 8.1. A particle in a box.

But in order that the probability of finding the particle outside the box be zero, $\psi^*\psi$ must be zero outside the box. Moreover if the wave function is to be continuous, $\psi^*\psi$ must also be zero at the walls of the box and ψ itself must be zero at the walls. Thus the two *boundary conditions* are :

$$\psi = 0 \text{ when } x = 0, \tag{8.14}$$

$$\psi = 0 \text{ when } x = a. \tag{8.15}$$

In addition to satisfying the Schrodinger amplitude equation, an acceptable amplitude function must *also* satisfy each of the imposed boundary conditions. Let's then apply the first boundary condition to ψ_1 as given by

$$\psi_1 = A \exp\left[\frac{i2\pi}{h}(2mE)^{1/2}x\right].$$

When $x = 0$, then $\psi_1 = 0$, so that

$$0 = A \exp(0) = A.$$

But A cannot be zero, because if it were, ψ_1 as given by Eq. (8.3) would *always* be equal to zero, and $\psi_1^*\psi_1$ would also always be zero at *any* value of x, so that the particle would have a zero probability of existing anywhere! In other words, if $A = 0$, the particle cannot exist in the box! A similar attempt to apply the first boundary condition to ψ_2 as given by Eq. (8.4) leads to the equally impossible conclusion that $B = 0$. Thus, although ψ_1 or ψ_2 is each individually a satisfactory solution to the Schrodinger amplitude equation, neither is able to satisfactorily incorporate the first boundary condition.

We can solve the problem by constructing a new solution through linear combination. It has already been proved that if ψ_1 and ψ_2 is each a satisfactory solution to the Schrodinger amplitude equation and if each of the functions corresponds to the same eigenvalue energy, then *any* linear combination of ψ_1 and ψ_2 is also a satisfactory solution. We may write ψ which is a proposed new solution, as

$$\psi = \psi_1 + \psi_2, \tag{8.16}$$

or

$$\psi = A \exp\left[\frac{i2\pi}{h}(2mE)^{1/2}x\right] + B \exp\left[-\frac{i2\pi}{h}(2mE)^{1/2}x\right]. \tag{8.17}$$

Equation (8.17) is identical to Eq. (8.12), which we have applied to the motion in *either* direction of a free particle. The physical significance of Eq. (8.17) is as follows

We must allow the particle to move in either direction if we are to keep it in the box.

We have already seen that if *either* A or B is zero, so that motion of the particle is restricted to one direction only, the probability of finding the particle in the box is zero. Now let's apply the first boundary condition to Eq. (8.17), that is, $\psi = 0$ when $x = 0$. Substitution of $x = 0$ and $\psi = 0$ in Eq. (8.17) yields

$$0 = A + B,$$

or

$$B = -A. \tag{8.18}$$

Then Eq. (8.17) becomes

$$\psi = A\left\{\exp\left[\frac{i2\pi}{h}(2mE)^{1/2}x\right] - \exp\left[-\frac{i2\pi}{h}(2mE)^{1/2}x\right]\right\}, \quad (8.19)$$

a function which now satisfies both the Schrodinger amplitude equation *and* the first boundary condition. If we assume $y = 2\pi(2mE)^{1/2}x/h$, we may write Eq. (8.19) as

$$\psi = A(e^{iy} - e^{-iy}). \quad (8.20)$$

But, according to Euler's formula,

$$e^{iy} = \cos y + i \sin y \text{ and } e^{-iy} = \cos y - i \sin y,$$

so that

$$e^{iy} - e^{-iy} = 2i \sin y, \quad (8.21)$$

and Eq. (8.20) becomes, after substitution for y,

$$\psi = i2A \sin\left[\frac{2\pi}{h}(2mE)^{1/2}x\right] = C \sin\left[\frac{2\pi}{h}(2mE)^{1/2}x\right], \quad (8.22)$$

where $C = i2A$ (an imaginary number only if A is real).

Let us now impose the second boundary condition, that is, $\psi = 0$ when $x = a$. Substitution of $\psi = 0$ and $x = a$ into Eq. (8.22) yields

$$0 = C \sin\left[\frac{2\pi}{h}(2mE)^{1/2}a\right]. \quad (8.23)$$

Since $C \neq 0$, $\sin\,[2\pi(2mE)^{1/2}a/h]$ must be equal to zero, which is possible whenever the argument of the sine is an integral multiple of π, that is, when

$$\frac{2\pi}{h}(2mE)^{1/2}a = \pm n\pi, \quad (8.24)$$

where $n = 0, 1, 2, 3, \ldots, \infty$.

From Eq. (8.24), then,

$$\frac{2}{h}(2mE)^{1/2} = \pm n/a. \quad (8.25)$$

Substitution of Eq. (8.25) into Eq. (8.22) yields

$$\psi = C \sin\left(\pm\frac{n\pi x}{a}\right) = \pm C \sin\left(\frac{n\pi x}{a}\right) \tag{8.26}$$

The amplitude functions ψ given by Eq. (8.26) now satisfy the Schrodinger amplitude equation and both boundary conditions. That is, when x is 0 or a, $\psi = 0$. Then $\pm$ sing in Eq. (8.26) indicates that there are two satisfactory series of functions, either the positive series or the negative series, each containing one function for each value of n.

Finally we have to evaluate C by applying the normalization condition, that is,

$$\int_0^a \psi^*\psi \, dx = 1. \tag{8.27}$$

The normalization condition ensures that each wave function is such that the total probability for finding the particles between $x = 0$ and $x = a$ (in the "box") is unity. From Eq. (8.26) we may write ψ^* as

$$\psi^* = \pm C^* \sin\left(\frac{n\pi x}{a}\right) \tag{8.28}$$

Substitution of *either* the positive or negative versions of Eqs. (8.25) and (8.28) into Eq. (8.27) yields

$$C^*C\int_0^a \sin^2\left(\frac{n\pi x}{a}\right) dx = 1. \tag{8.29}$$

Integration of Eq. (8.29) yields

$$C^*C\left[\frac{x}{2} - \frac{\sin(2\pi n x/a)}{4\pi n/a}\right]_0^a = 1,$$

or

$$C^*C\left(\frac{a}{2}\right) = 1 \text{ or } |C|^2 = \frac{2}{a}.$$

When C is expressed as a real number, Eq. (8.26) becomes

$$\psi = \pm\left(\frac{2}{a}\right)^{1/2} \sin\left(\frac{n\pi x}{a}\right) \tag{8.30}$$

However, because probability is given by $|\psi|^2$ rather than by ψ, it is immaterial whether C is chosen as the positive or negative square root.

This means that it is not necessary to carry both a positive *and* a negative function for each value of n, since *either* the positive or the negative function alone can be used to calculate probabilities and to generate exact and expectation values for desired observables, including energy levels, for any given value of n. Thus, we discard the set of negative functions and adopt the positive series of normalized functions for the single particle in the one-dimensional box as

$$\psi = \left(\frac{2}{a}\right)^{1/2} \sin\left(\frac{n\pi x}{a}\right) \tag{8.31}$$

where $n = 1, 2, 3, \ldots, \infty$. Note that we have eliminated the solution for which $n = 0$, since when $n = 0$, then $\psi = 0$ and $\psi^*\psi = 0$ everywhere. That is, when $n = 0$, *the particle does not exist.* According to Eq. (8.31), there is an *infinite number* of satisfactory discrete functions, each involving a different integral value for n, the *quantum number.*

ALLOWED ENERGY LEVELS AND PROBABILITY DISTRIBUTION FOR A PARTICLE IN A ONE-DIMENSIONAL BOX

The allowed values for the energy of the particle in the one-dimensional box are given by squaring and rearranging Eq. (8.25), which yields,

$$E = n^2h^2/8ma^2, \tag{8.32}$$

where $n = 1, 2, 3, \ldots, \infty$. Equation (8.32) expresses the stationary-state energies available to the particle in the box. *No other exact energies are allowed nor can other exact energies be experimentally observed.*

The state of the particle characterized by a particular value for one of the allowed energy levels is known as *a quantum state.* For example, in the first *quantum state,* or *ground state,* $n = 1$, and according to Eq. (8.32) the energy is,

$$E_1 = h^2/8ma^2.$$

It is important to note that the allowed energy states given by Eq. (8.32) could also have been easily generated by the operation

$$\hat{T}\psi = T\psi, \tag{8.33}$$

according to the second postulate, since $E = T$. While using Eq. (8.33), $\hat{T} = (-h^2/8\pi^2m)/(d^2/d\,x^2)$, and the ψ functions are given by Eq. (8.31).

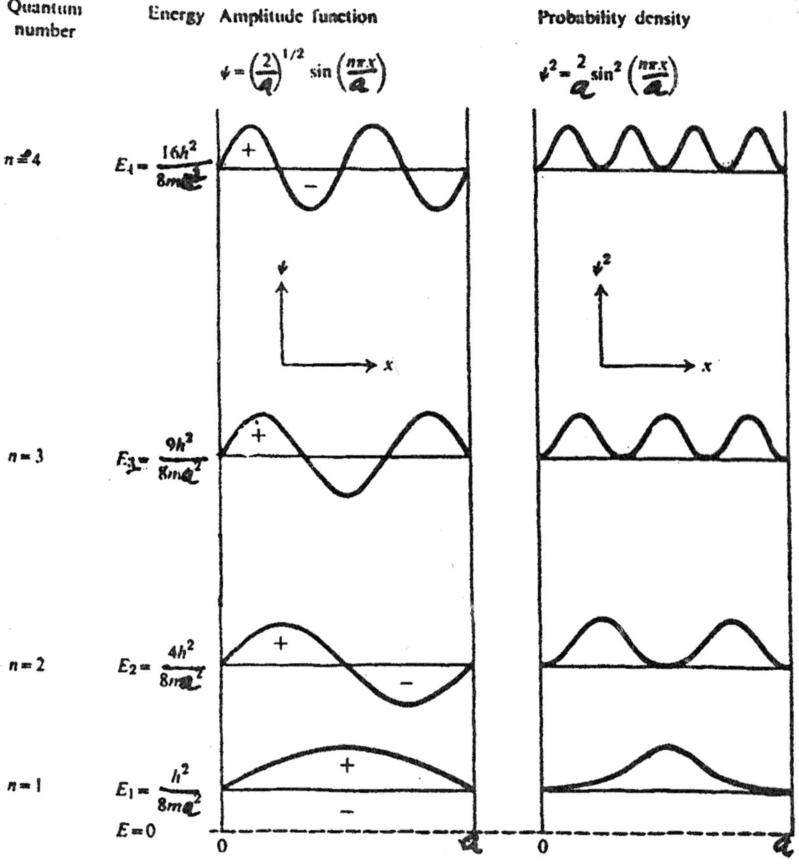

Fig. 8.2. Energy levels, amplitude functions, and probability densities for the single particle in the one-dimensional box.

In order to find out where the particle might be localized within the one- dimensional box we'll now develop the expression for $\psi^*\psi$, the probability density, as a function of x. The complex conjugate of ψ is equal to ψ, since i does not appear in the equation for ψ. Thus, for any value of n,

$$\psi^*\psi = |\psi|^2 = \psi^2 = \frac{2}{a}\sin^2\left(\frac{n\pi x}{a}\right) \tag{8.34}$$

where $n = 1, 2, 3, \ldots, \infty$. According to Eq. (8.34), the probability density is a function of both x and n.

The allowed amplitude functions for various values of the quantum number n are shown in Fig. 8.2. Corresponding energy levels and probability distributions are also shown. As for the vibrating string fixed at both ends, n is equal to the number of antinodes or loops. The energy of a quantum state increases as the square of n. In the case of the fixed vibrating string we have already noted that the energy in the string was higher for higher values of n. Since in a system of fixed total energy higher-energy levels are statistically less probable at thermal equilibrium. It is expected that in a system consisting of a large number of particles at equilibrium, distributed over all of the allowed quantum states, most of the particles would be in the ground state, for which $n = 1$.

The magnitude of the energy levels, and thus the magnitude of the separation between energy levels, is inversely proportional to ma^2. Thus, for large masses or large values of a (large "boxes"), the allowed energy levels are very close together and appear to be continuous. For such systems, quantum mechanics reduces to classical mechanics. When m and a are small, however, the energy in the second allowed energy state may be appreciably above the energy in the ground state and the energy difference may become experimentally detectable. Clarify the difference between "classical" and "quantum" systems.

Example 8.1 : The following example consider a 10-g marble in a 100-cm one-dimensional box. According to Eq. (8.32), the lowest energy the marble may have is that energy for which $n = 1$ (ground-state energy), which is

$$E_1 = \frac{n^2h^2}{8ma^2} = \frac{(1)^2(6.63 \times 10^{-27})^2}{8(10)(100)^2}$$

$$= 5.48 \times 10^{-59} \text{ erg.}$$

The next highest allowed energy level is

$$E_2 = (2)^2E_1 = 21.9 \times 10^{-59} \text{ erg.}$$

If the marble were to drop in energy from the first excited state E_2 to the ground state E_1, the energy lost would be $\varepsilon = E_2 - E_1 = 16.4 \times 10^{-59}$ erg. In turn, if this minute amount of energy were to be emitted as a single photon, the wavelength of the emitted radiation, as calculated from the Planck-Einstein relationship

$$\varepsilon = h\nu = hc/\lambda, \tag{8.35}$$

is given by

$$\lambda = \frac{hc}{\varepsilon} = \frac{(6.63 \times 10^{-27})(3 \times 10^{10})}{16.4 \times 10^{-59}} = \sim 10^{42} \text{ cm},$$

and the corresponding period of the photon would be

$$\tau = \frac{1}{\nu} = \frac{h}{\varepsilon} = \frac{6.63 \times 10^{-27}}{16.4 \times 10^{-59}} = 4 \times 10^{31} \text{ sec}$$

$$= \sim 10^{2} \text{ years},$$

which is a long time to wait for a classical measurement! Because of the closeness in spacing between adjacent energy levels of the marble and the resulting limitations in our ability to measure the corresponding energy changes, the spectrum of energy states available to the marble appears experimentally as if it were continuous.

Example 8.2 : Let's calculate the first two allowed energy levels for a mass the size of an electron ($m = 9.1 \times 10^{-28}$ g) confined in a one-dimensional box about the size of an atom ($a = 3 \times 10^{-8}$ cm). The ground-state energy is

$$E_1 = \frac{n^2h^2}{8ma^2} = \frac{(1)^2(6.63 \times 10^{-27})^2}{8(9.1 \times 10^{-23})(3 \times 10^{-8})^2} = 6.7 \times 10^{-12} \text{ erg},$$

and the energy in the second energy level is

$$E_2 = (2)^2E_1 = 26.8 \times 10^{-12} \text{ erg}.$$

The energy evolved when the electron undergoes a transition from E_2 to E_1 is given as

$$\varepsilon = E_2 - E_1 = 20.1 \times 10^{-12} \text{ erg}.$$

If this amount of energy is emitted as a single photon, the wavelength of the emitted radiation is

$$\lambda = \frac{hc}{\varepsilon} = \frac{(6.63 \times 10^{-27})(3 \times 10^{10})}{20.1 \times 10^{-12}} = 9.9 \times 10^{-6}\, cm = 990 \text{ Å}.$$

The radiation emitted is a high-energy photon in the ultraviolet region. This can be detected quite easily experimentally. In general, we can measure readily most differences in allowed electron energies within atoms.

PROBABILITY AND PROBABILITY DENSITY

Consider the probability distribution shown in Fig. 8.2. In each state, the probability of finding the particle at the wall is zero. This is expected as per the defined boundary conditions. However, when n

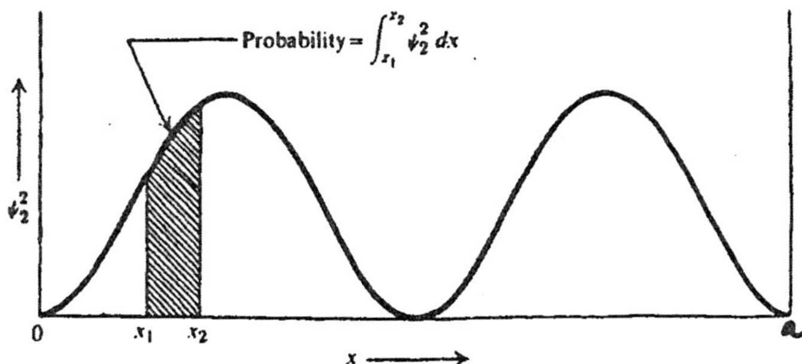

Fig. 8.3. A graphical representation of the probability that a single particle exists between x_1 and x_2 in a one- dimensional box. ψ_2^2 is the probability density for the second allowed quantum state (n = 2).

is greater than one, there are points of zero probability for finding the electron at values for x other than 0 or a. For example, for the second allowed state, in which $n = 2$, the probability of finding the electron at $x = a/2$ is zero and we experience maximum probability densities at $x = a/4$ and $x = 3a/4$. Consider the $n = 2$ state in more detail to understand the meaning of the terms "probability" and "probability density." The curve for $|\psi_2|^2$ versus x is shown in Fig. 8.3. The probability that the particle exists between any two positions x_1 and x_2, is given by the expression

$$\int_{x_1}^{x_2} |\psi_2|^2\, dx = \int_{x_1}^{x_2} \psi_2^2\, dx, \tag{8.36}$$

where $|\psi_2|^2$ is the probability density, or probability *per unit length of x*. Graphically, the probability is simply the area under the curve between x_1 and x_2 (the shaded area in Fig. 8.3). Normalization of ψ ensures that the *total* area under the curve from $x = 0$ to $x = a$ is unity, so that the total probability is unity. We are thus sure to find the particle somewhere in the box.

If the single particle has a finite probability of existing between $x = 0$ and $x = l/2$ and also has a finite probability of existing between $x = a/2$ and $x = a$, but has zero probability of existing at the position $x = a/2$, how does it get from one side of the box to the other? The answer to this question lies in the more refined *relativistic* treatment of Dirac, which is mathematically much more complicated than the nonrelativistic Schrodinger treatment (used here). When the problem of the particle in the box is solved by Dirac's relativistic approach, one of the important consequences is the *disappearance* of the nodes of zero probability. In place of the nodes there result *approximate nodes* at which the probability density, although very small, is nevertheless *finite.* This implies that the single particle has a small but finite probability for existing at the center of the box, and may, therefore, pass from one side to the other. Similarly even though nonrelativistic treatment of the electron in the hydrogen atom will lead to nodes, or surfaces in space at which the probability density for finding the electron is zero, it is important to realize that, in terms of the more correct relativistic treatment, the "Schrodinger nodes" are not true nodes at all, but rather are approximate nodes which represent regions of very small but finite probability density.

ORTHOGONALITY OF WAVE FUNCTIONS

For the single particle in one-dimensional box there is a series of acceptance wave functions generated through the second Postulate (Eq. 8.13), each of which is an eigenfunction of the Hamiltonian operator $\hat{H}$ and each of which conforms to appropriate boundary conditions. These are given by Eq. (8.21). Each of the following is an acceptable wave function :

$$\psi_1 = \left(\frac{2}{a}\right)^{1/2} \sin\frac{\pi x}{a},\ \psi_2 = \left(\frac{2}{a}\right)^{1/2} \sin\frac{2\pi x}{a},\ \psi_3 = \left(\frac{2}{a}\right)^{1/2} \sin\frac{3\pi x}{a},$$

and so on.

Each of the generated eigenfunctions is, in a special way, *independent* of any other. This independence is expressed in a property called *orthogonality*.

By definitions, two wave functions, ψ_m and ψ_n, where m and n are two different quantum number, are *orthogonal* if

$$\int_{\text{all space}} \psi_m^* \psi_n \, d\tau = 0. \tag{8.37}$$

For the single particle in the one-dimensional box, Eq. (8.37) may be written as

$$\int_0^a \psi_m^* \psi_n \, dx = 0, \tag{8.38}$$

where

$$\psi_m^* = \psi_m = \left(\frac{2}{a}\right)^{1/2} \sin \frac{m\pi x}{a}, \tag{8.39}$$

$$\psi_n = \left(\frac{2}{a}\right)^{1/2} \sin \frac{m\pi x}{a}. \tag{8.40}$$

It is easily proved that any two of the generated wave functions for the particle in the one-dimensional box are orthogonal by substitution of Eqs. (8.39) and (8.40) into Eq. (8.38), which gives

$$\int_0^a \psi_m^* \psi_n \, dx = \frac{2}{a} \int_0^a \sin \frac{m\pi x}{a} \sin \frac{n\pi x}{a} \, dx. \tag{8.41}$$

But one of the fundamental trigonometric identities is

$$\sin \alpha \cdot \sin \beta = \frac{1}{2} \cos (\alpha - \beta) - \frac{1}{2} \cos (\alpha - \beta).$$

Therefore Eq. (8.41) may be expressed as :

$$\int_0^a \psi_m^* \psi_n \, dx = \frac{1}{a} \int_0^a \left[\cos (m - n) \pi \frac{x}{a} - \cos (m + n) \pi \frac{x}{a}\right] dx$$

which, in $m \neq n$, may be integrated as

$$\frac{1}{a}\left[\frac{a}{(m - n) \pi} \cdot \sin (m - n) \pi \frac{x}{a} - \frac{a}{(m + n) \pi} \cdot \sin (m + n) \pi \frac{x}{a}\right]_0^a = 0,$$

because $m - n$ are integers.

Since any one of the wave functions for the particle in the one-dimensional box is orthogonal to any other of the wave functions, the entire set of solutions is referred to as an ***orthogonal set.*** Moreover since each wave function in the set is not only orthogonal to all others, but is also normalized, the set of solutions given by Eq. (8.31) is called an *orthogonal* set (***orthogonal*** and ***normalized***).

In order to describe certain physical systems, it is convenient to construct new wave functions through linear combinations of orthogonal eigenfunctions. It is important to recognize that such linear-combination functions are not *generally* orthogonal. The following example illustrate the manipulation : Consider the two wave functions ψ_{I} and ψ_{II} generated by arbitrary linear combinations of the orthogonal eigenfunctions ψ_m and ψ_n for the particle in the one-dimensional box,

$$\psi_{\text{I}} = a\psi_m + b\psi_n, \ \psi_{\text{II}} = c\psi_m + d\psi_n,$$

where a, b, c and d are constants to be determined.

Although either of the two linear-combination functions ψ_{I} or ψ_{II} may under certain circumstances represent a *physical satisfactory description* of a system, note that neither ψ_{I} nor ψ_{II} is an eigenfunction of the Hamiltonian operator for the particle in the one-dimensional box unless the eigenvalue energies E_m and E_n are identical. In order for ψ_{I} and ψ_{II} to be orthogonal, the condition is

$$\int_0^L \psi_{\text{I}}^* \psi_{\text{II}} \, dx = 0.$$

But since all of the eigenfunctions for the particle in the one- dimensional box are real, $\psi_{\text{I}}^* = \psi_{\text{I}}$,

$$\int_0^L \psi_{\text{I}} \psi_{\text{II}} \, dx = 0, \ \int_0^L (a\psi_m + b\psi_n)(c\psi_m + d\psi_n) \, dx = 0.$$

Multiplication of terms, followed by rearrangement, gives

$$ac\int_0^L \psi_m\psi_m \, dx + ad\int_0^L \psi_n\psi_m \, dx + bc\int_0^L \psi_n\psi_n \, dx + bd\int_0^L \psi_n\psi_n \, dx = 0. \tag{(8.42)}$$

Since ψ_m and ψ_n are orthogonal, the integrals in the second and third terms in Eq. (8.42) are zero, and since ψ_m and ψ_n are each normalized, the integrals in the first and fourth terms are unity, so that

$$ac + bd = 0,$$

which is requirement if ψ_{I} and ψ_{II} are to be orthogonal. It can be easily shown, for example, that if $a = b$ and $c = -d$ the linear combination functions

$$\psi_{\text{I}} = a(\psi_m + \psi_n),\ \psi_{\text{II}} = c(\psi_m - \psi_n),$$

where a or c may assume any constant value, are orthogonal in the interval from $x = 0$ to $x = L$. If, in addition, ψ_{I} and ψ_{II} are also to be *normalized*, it may be established that a and c each must be equal to $1/\sqrt{2}$, so that

$$\psi_{\text{I}} = \frac{1}{\sqrt{2}}(\psi_m + \psi_n) \text{ and } \psi_{\text{I}} = \frac{1}{\sqrt{2}}(\psi_m - \psi_n) \qquad (8.43)$$

are now *orthonormalized.*

It can be easily proved that linear-combination functions which do *not* contain any orthogonal eigenfunction in common are *always* orthogonal, and that linear-combination functions which contain *only one* orthogonal eigenfunction in common can never be made orthogonal.

PARTICLE IN A THREE-DIMENSIONAL BOX

Let's now consider a single particle confined in a three- dimensional box of dimensions a, b, and c in the x-, y-, and z-directions, respectively. In order to ensure that the particle remains in the box, the potential energy V must be zero everywhere within the box and infinite everywhere outside the box. Then within the box, where $V = 0$, the Schrodinger amplitude equation may be written as

$$\frac{\partial^2\psi(x, y, z)}{\partial x^2} + \frac{\partial^2\psi(x, y, z)}{\partial y^2} + \frac{\partial^2\psi(x, y, z)}{\partial z^2} = \frac{-8\pi^2 mE}{h^2}\psi(x, y, z). \quad (8.44)$$

Equation (8.44) cannot be solved as it is, because of its dependence on three variables. It is possible to separate the variables in similar linear differential equations by assuming that the state function may be expressed as a *product* of several functions, each function dependent on one of the variables. Let's assume then that

$$\psi(x, y, z) = \varphi(x)\varphi(y)\varphi(z), \tag{8.45}$$

where $\varphi(x)$ depends only on x, $\varphi(y)$ depednds only on y, and $\varphi(z)$ depends only on z. In order to determine the satisfactory forms for $\varphi(x)$, $\varphi(y)$, and $\varphi(z)$, let us substitute Eq. (8.45) into Eq. (8.44) :

$$\varphi(y)\varphi(z)\frac{d^2\varphi(x)}{dx^2} + \varphi(x)\varphi(z)\frac{d^2\varphi(y)}{dy^2}$$

$$+ \varphi(x)\varphi(y)\frac{d^2\varphi(z)}{dz^2} = \frac{-8\pi^2 mE}{h^2}\varphi(x)\varphi(y)\varphi(z). \tag{8.46}$$

Note that the derivatives in Eq. (8.46) are now expressed as *total* derivatives rather than partial derivatives, sineteach of the φ-functions is dependent on only one of the coordinate variables. Dividing Eq. (8.46) by $\varphi(x)\varphi(y)\varphi(z)$ we have

$$\frac{1}{\varphi(x)}\cdot\frac{d^2\varphi(x)}{dx^2} + \frac{1}{\varphi(y)}\cdot\frac{d^2\varphi(y)}{dy^2} + \frac{1}{\varphi(z)}\cdot\frac{d^2\varphi(z)}{dz^2} = \frac{-8\pi^2 mE}{h^2}. \tag{8.47}$$

Since E is constant in a conservative force field ($E = T$ = constant in this specific case), the sum of the three terms on the left-hand side of Eq. 98.47) is constant irrespective of particular values of x, y, or z. Moreover, a change in the value of x cannot change the value of either the second or the third term on the left-hand side of Eq. (8.47), since each is independent of x. Thus the first term itself must be constant. By a similar argument the second term and the third term on the left side are also constants. The total energy (kinetic) of the particle in terms of the component energies in the three coordinate directions may be expressed as

$$E = E_x + E_y + E_z = \text{constant}. \tag{8.48}$$

Substitution of Eq. (8.48) into Eq. (8.47) gives

$$\frac{1}{\varphi(x)}\frac{d^2\varphi(x)}{dx^2} + \frac{1}{\varphi(y)}\frac{d^2\varphi(y)}{dy^2} + \frac{1}{\varphi(z)}\frac{d^2\varphi(z)}{dz^2}$$

$$= \left(\frac{-8\pi^2 mE_x}{h^2}\right) + \left(\frac{-8\pi^2 mE_y}{h^2}\right) + \left(\frac{-8\pi^2 mE_z}{h^2}\right) \tag{8.49}$$

Since each of the terms on the left side of Eq. (8.49) is separately equal to a constant, the three independent equations may be expressed as follows :

$$\frac{1}{\varphi(x)}\frac{d^2\varphi(x)}{dx^2}=\frac{-8\pi^2 mE_x}{h^2},$$

$$\frac{1}{\varphi(y)}\frac{d^2\varphi(y)}{dy^2}=\frac{-8\pi^2 mE_y}{h^2},$$

$$\frac{1}{\varphi(z)}\frac{d^2\varphi(z)}{dz^2}=\frac{-8\pi^2 mE_z}{h^2},$$

Rearranging to the more familiar forms, we have

$$\frac{d^2\varphi(x)}{dx^2}=\frac{-8\pi^2 mE_x}{h^2}\varphi(x), \tag{8.50}$$

$$\frac{d^2\varphi(y)}{dx^2}=\frac{-8\pi^2 mE_x}{h^2}\varphi(y), \tag{8.51}$$

$$\frac{d^2\varphi(z)}{dx^2}=\frac{-8\pi^2 mE_x}{h^2}\varphi(z), \tag{8.52}$$

Equation (8.50), (8.51), and (8.52) are one-dimensional eigenvalues equations identical to that for the particle in the one-dimensional box (Eq. 8.13). The individual normalized eigenfunctions are thus identical in form to Eq. (8.31) and may be written as

$$\varphi(x)=\left(\frac{2}{a}\right)^{1/2}\sin\frac{n_x\pi x}{a}. \tag{8.53}$$

$$\varphi(y)=\left(\frac{2}{a}\right)^{1/2}\sin\frac{n_y\pi yx}{b}. \tag{8.54}$$

$$\varphi(z)=\left(\frac{2}{a}\right)^{1/2}\sin\frac{n_z\pi z}{c}. \tag{8.55}$$

where n_x, n_y, and n_z are the quantum numbers in each of the coordinate directions such that

$$n_x = 1,\ 2,\ 3,\ \ldots,\ \infty,$$

$$n_y = 1,\ 2,\ 3,\ \ldots,\ \infty,$$

$$n_z = 1,\ 2,\ 3,\ \ldots,\ \infty,$$

Substitution of Eqs. (8.53), (8.54), and (8.55) into Eq. (8.45) yields as the amplitude function for the particle in the three- dimensional box

$$\psi(x, y, z) = \left(\frac{2}{a}\right)^{1/2}\left(\frac{2}{b}\right)^{1/2}\left(\frac{2}{c}\right)^{1/2} \sin\frac{n_x\pi x}{a} \sin\frac{n_y\pi y}{b} \sin\frac{n_z\pi z}{c}, \quad (8.56)$$

where n_x, n_y, and n_z may each have independently any integral value other than zero. If any value of n were zero, $\psi(x, y, z)$ would always be zero and the particle could not exist, since $\psi^*\psi$ would *always* be zero. Since $\varphi(x)$, $\varphi(y)$; and $\varphi(z)$ are each individually normalized within their respective one-dimensional limits, it follows that $\psi(x, y, z)$ as given by Eq. (8.56) must also be normalized within the dimensions of the box.

According to Eq. (8.32) it also follows that the allowed energy levels for the single particle in the three-dimensional box are given by

$$E + E_x + E_y + E_z = \frac{h^2 n_x^2}{8ma^2} + \frac{h^2 n_x^2}{8mb^2} + \frac{h^2 n_z^2}{8mc^2}, \quad (8.57)$$

or

$$E = \frac{h^2}{8m}\left(\frac{n_x^2}{a^2} + \frac{n_y^2}{b^2} + \frac{n_z^2}{c^2}\right) \quad (8.58)$$

ALLOWED ENERGY LEVELS IN A CUBIC BOX

If the box is cubic and each edge is equal to a, then Eq. (8.58) becomes

$$E = \frac{h^2}{8ma^2}(n_x^2 + n_y^2 + n_z^2). \quad (8.59)$$

The lowest allowed energy state, or *ground state,* is that state for which each of the three quantum numbers is equal to one, that is, $n_x = 1, n_y = 1$, and $n_z = 1$. We will refer to this state as the (111) state. Then, according to Eq. (8.59),

$$E_{(111)} = 3h^2/8ma^2.$$

The next highest energy state is state in which any one of the quantum numbers is two but the other two quantum numbers are each one. There are three such states :

(211) state : $n_x = 2,\ n_y = 1,\ n_z = 1,$

(121) state : $n_x = 1,\ n_y = 2,\ n_z = 1,$

$$(112) \text{ state} : n_x = 1,\ n_y = 1,\ n_z = 2.$$

For each of the above states, the energy calculated from Eq. (8.59) is identical. That is,

$$E_{(211)} = E_{(121)} = E_{(112)} = 6h^2/8ma^2.$$

The existence of several distinct quantum states having the same energy level has earlier been referred to as *degeneracy*. Thus, the energy level $6h^2/8ma^2$ is said to be *triply degenerate* or *threefold degenerate*

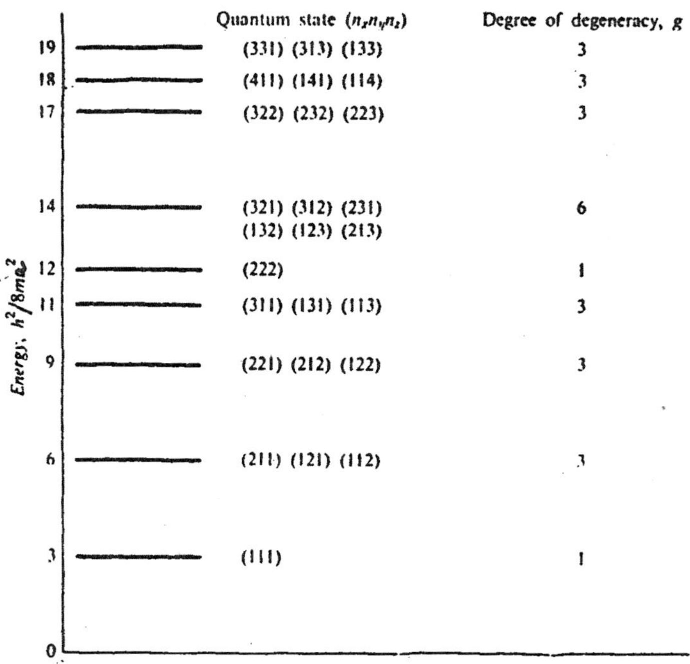

Fig. 8.4. Allowed energy levels for the particle in a cubic box.

or is said to have a *multiplicity* of three, or a *degree of degneracy* of three, or a *quantum weight* of three, or a *statistical weight* of three. In addition, such a state is often called a *triplet* state. The term *degree of degeneracy* is commonly used and this term has the symbol g. Thus, for the energy level $6h^2/8ma^2$, three degenerate quantum states, (211), (121), and (112), exist, and $g = 3$. Higher energy levels, corresponding quantum states, and corresponding degrees of degeneracy are given in Fig. 8.4 for the single particle in a cubic box.

It is interesting to note that the degeneracy we observe for the higher quantum states appears by virtue of the symmetry of the cube. If we had assumed $a \neq b \neq c$ in calculating energy levels from Eq. (8.58), such degeneracy would not occur. If we were to *distort* the cube, for instance, by pushing one wall in slightly, levels which were previously degenerate would split into separate energy levels. This splitting of degenerate electronic energy levels into new separate levels may be observed if we distort atomic symmetry by applying a strong directional electrical or magnetic field to a symmetrical atom. The corresponding loss of degeneracy results in new lines in the emission spectrum of the atom. The splitting of lines in atomic spectra in the presence of a directional magnetic field is known as the *Zeeman effect;* and the splitting of lines in the presence of a directional electrical field is known as the *Stark effect.*

PROBABILITY DENSITIES FOR A PARTICLE IN A CUBIC BOX

For a cubic box of edge a, Eq. (8.56) becomes

$$\psi(x, y, z) = \left(\frac{2}{a}\right)^{3/2} \sin\frac{n_x \pi x}{a} \sin\frac{n_y \pi y}{a} \sin\frac{n_z \pi z}{a}, \tag{8.60}$$

and since

$$\psi(x, y, z) = \varphi(x)\varphi(y)\varphi(z),$$

we may write the contributing one-dimensional functions as follows :

$$\varphi(x) = \left(\frac{2}{a}\right)^{1/2} \sin\frac{n_x \pi x}{a}, \tag{8.61}$$

$$\varphi(y) = \left(\frac{2}{a}\right)^{1/2} \sin\frac{n_y \pi y}{a}, \tag{8.62}$$

$$\varphi(z) = \left(\frac{2}{a}\right)^{1/2} \sin\frac{n_z \pi z}{a}, \tag{8.63}$$

Amplitude functions in each of the coordinate directions are plotted for the (111) state in Fig. 8.5. The corresponding probability densities in each of the coordinate directions are

$$\varphi(x)^*\varphi(x) = \varphi^2(x) = \frac{2}{a} \sin^2\frac{n_x \pi x}{a}, \tag{8.64}$$

$$\varphi(y)^*\varphi(y) = \varphi^2(y) = \frac{2}{a}\sin^2\frac{n_y\pi y}{a}, \tag{8.65}$$

$$\varphi(z)^*\varphi(z) = \varphi^2(z) = \frac{2}{a}\sin^2\frac{n_z\pi z}{a}, \tag{8.66}$$

Directional probability densities for the (111) quantum state are also plotted in Fig. 8.5. The φ^2-values represent the probability *per unit length* that the particle exists at a given coordinate position.

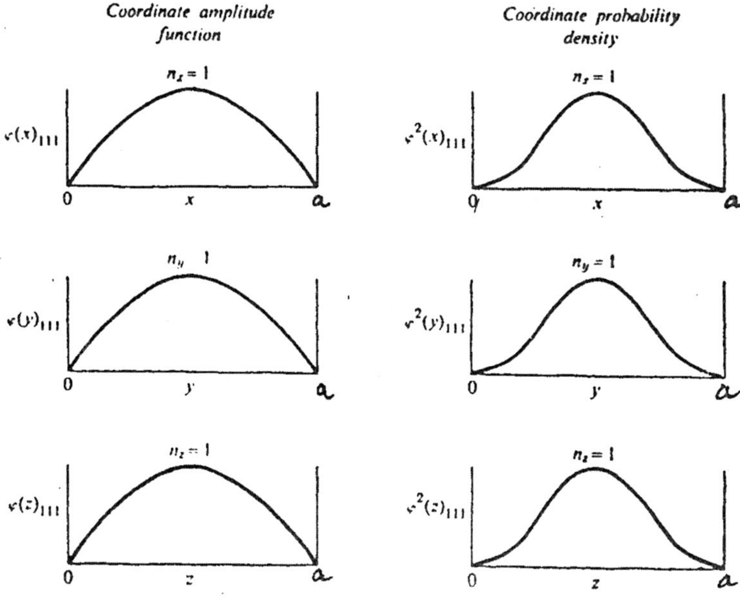

Fig. 8.5. Amplitude functions and probability densities for the *x*, *y*, and *z* coordinate directions for the (111) quantum state of the particle in a cubic box.

The probability density for the particle in three dimensional is given by the relationship

$$\psi^*\psi = \varphi(x)^*\varphi(x)\varphi(y)^*\varphi(y)\varphi(z)^*\varphi(z),$$

or

$$\psi^2 = \varphi^2(x)\cdot\varphi^2(y)\cdot\varphi^2(z), \tag{8.67}$$

which is the probability per unit *volume* that the particle exists at the point *x*, *y*, *z*. Graphical multiplication of the φ- and φ^2-probability function given for the (111) state in Fig. 8.5 leads to the three- dimensional representations for ψ and ψ^2 shown in Fig. 8.6. In the three-dimensional plots the density of the amplitude cloud or of the probability cloud represents a direct measure of ψ or ψ^2, respectively. Also note from Fig. 8.6 that the particle has the highest probability density at the center of the cube and has zero probability density at the walls. Thus each of the walls is a *nodal plane.*

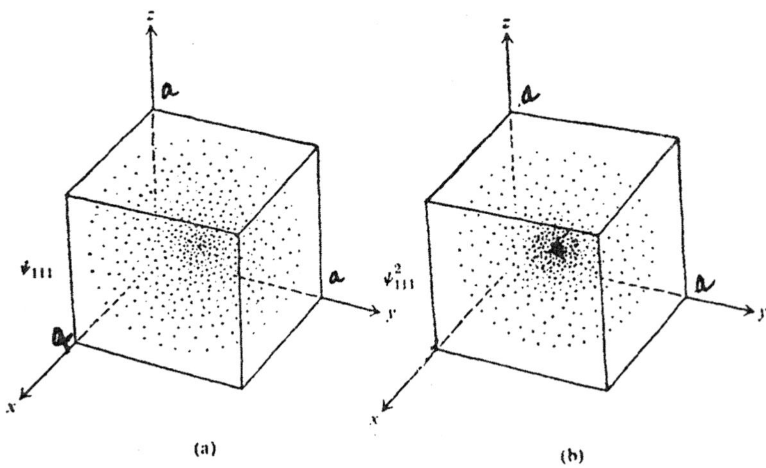

Fig. 8.6. Amplitude function ψ and probability density ψ^2 for the particle in a cubic box in the (111) state. The density of the cloud is proportional to the value of ψ or ψ^2 at a particular coordinate position.

A contour plot of a cross section of the three-dimensional ψ^2-plot of Fig. 8.6 through the plane $x = a/2$ is shown in Fig. 8.7.

Since

$$\psi^2 = \frac{8}{a^3} \sin^2\left(\frac{n_x \pi x}{a}\right) \sin^2\left(\frac{n_y \pi y}{a}\right) \sin^2\left(\frac{n_z \pi z}{a}\right),$$

the probability density ψ^2 has its maximum value $8/a^3$ when each of the $\sin^2$ terms is equal to 1. In the (111) state this maximum occurs at $x = a/2$, $y = a/2$, and $z = a/2$. Thus for convenience the contour plot in Fig. 8.7 presents ψ^2-values in units of $8/a^3$. The contour lines represent different constant values for ψ^2. In the case of the symmetrical (111) state, contours in the planes for which $y = a/2$ and $z = a/2$ would be identical to those in the $(x = a/2)$-plane.

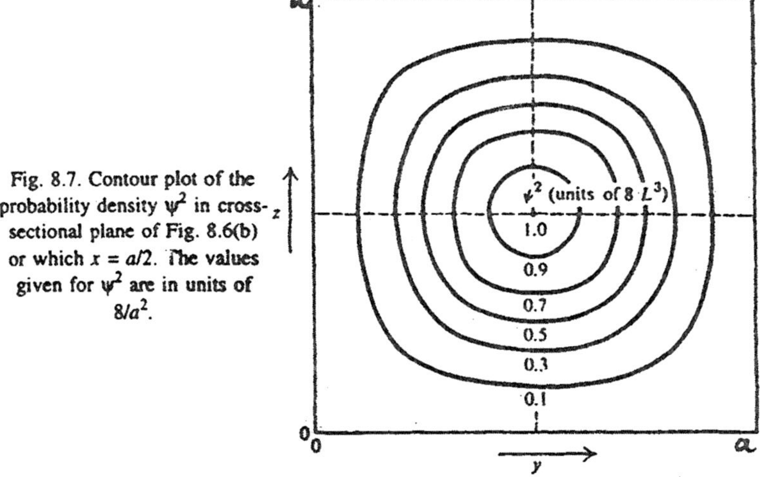

Fig. 8.7. Contour plot of the probability density ψ^2 in cross-sectional plane of Fig. 8.6(b) or which $x = a/2$. The values given for ψ^2 are in units of $8/a^2$.

In pictorializing atomic and molecular orbitals, it is often convenient to represent three-dimensional ψ- and ψ^2-plots by drawing the boundary surface which encloses all points having ψ- or ψ^2- values above some relatively small value such as 0.10 or 0.02 of the maximum. Thus in order to present the general shape of the ψ^2-plot for the (111) state, we may draw the three- dimensional boundary surface which encloses all points for which ψ^2 (in units of $8/a^3$) has a value

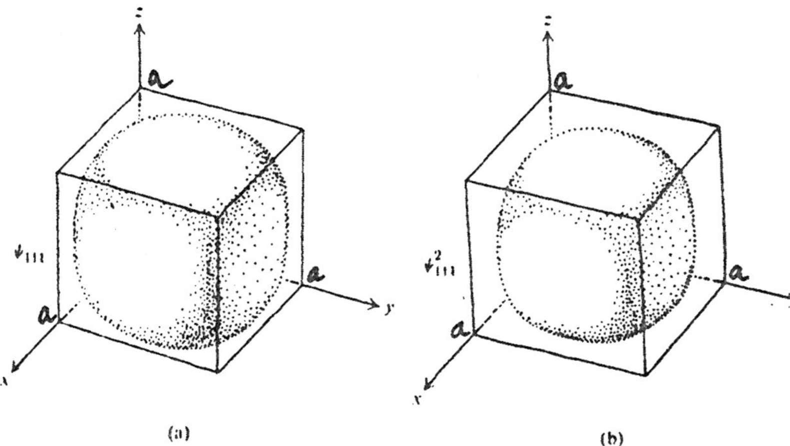

Fig. 8.8. Boundary surfaces for ψ and ψ^2 for the (111) state of the particle in a cubic box. Nodal planes occur at each of the walls.

larger than 0.10. Such boundary surface plots for ψ and for ψ^2 are shown in Fig. 8.8.

FINITE BARRIERS

The general initial forms of the wave function for a particle in one dimension and of the wave function for the particle in a one-dimensional box are identical and are given by slight rearrangement of Eq. (8.17) as

$$\psi = A \exp (i(8\pi^2 mE/h^2)^{1/2}x] + B \exp [-i(8\pi^2 mE/h^2)^{1/2}x], \quad (8.68)$$

in which E is constant for conservative systems. If we define

$$\alpha^2 \equiv 8\pi^2 mE/h^2, \quad (8.69)$$

we may write Eq. (8.68) in the simpler form :

$$\psi = A \exp [i\alpha x] + B \exp [-i\alpha x]. \quad (8.70)$$

In the above equation the first term is the wave function for the particle traveling in the $(+x)$-direction, whereas the second term is the wave function for the particle traveling in the $(-x)$-direction. The probability density for the particle traveling in the $(+x)$-direction at a given coordinate position x is given as $|A|^2$. Similarly, $|B|^2$ is the probability density for the particle traveling in the $(-x)$-direction. For a free particle which is restricted to motion in either the $(+x)$-direction or the $(-x)$-direction *only*, either $|A|^2$ *or* $|B|^2$ must be equal to zero.

If we confine a particle within a one-dimensional box, it is necessary that $|A|^2$ be equal to $|B|^2$; otherwise the particle would have a finite probability of escaping from the box.

We will now consider a different variation of the particle in the box problem. Consider a single particle within a one-dimensional box in which the potential-energy barrier is infinite at the left end but *finite* and equal to V_0 at the right. Let the potential energy be zero everywhere within the box. Furthermore, let the finite potential-energy barrier have a *finite thickness a* such that the potential energy is again zero to the right of the barrier. The system is shown in Fig. 8.9. Let $x = 0$ at the left edge of the barrier.

Consider a conservative system, in which the total energy of the particle is constant and equal to E. From classical mechanics it is obvious that if E is less than V_0 the particle should not have sufficient energy to escape from the box (region I) and thus should have zero probability of existing in region III to the right of the finite barrier. The results of quantum mechanics, however, are different and we will see that *even if E is less than V_0 the single particle still has a finite probability for existence beyond and to the right of the finite barrier.*

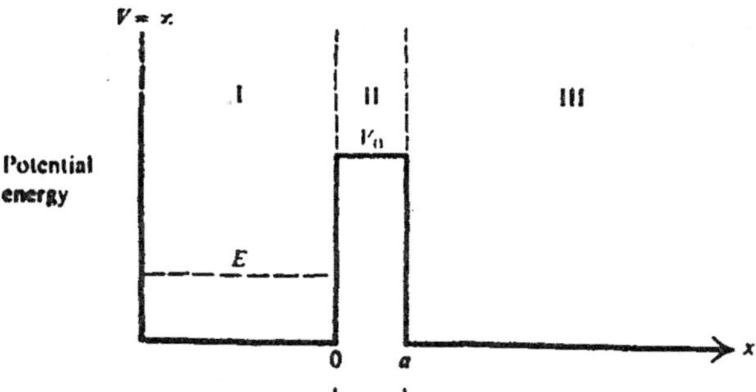

Fig. 8.9. Potential Energy as a function of x.

The Schrodinger equation (Eq. 8.1) may be rearranged as follows:

$$\frac{d^2\psi}{dx^2} = \frac{8\pi^2 m}{h^2}(V - E)\psi. \tag{8.71}$$

The Schrodinger equation for the three regions are as follows :

For region, I, where $V = 0$,

$$\frac{d^2\psi_I}{dx^2} + \alpha^2\psi_I = 0 \text{ (region I)},$$

where $\alpha^2 = 8\pi^2 mE/h^2$ according to Eq. (8.69).

For region II, where $V = V_0$, the Schrodinger equation may be written as

$$\frac{d^2\psi_{II}}{dx^2} - \beta^2\psi_{II} = 0 \text{ (region II)},$$

where

$$\beta^2 = \frac{8\pi^2 m(V_0 - E)}{h^2}. \tag{8.72}$$

Since $V = 0$ in region III, the Schrodinger equation for region III is identical to that for region I; that is,

$$\frac{d\psi_{III}}{dx^2} + \alpha^2\psi_{III} = 0 \text{ (region III)}.$$

The wave functions for the three regions are :

$$\psi_I = A \exp[i\alpha x] + B \exp[-i\alpha x], \tag{8.73}$$

$$\psi_{II} = C \exp[\beta x] + D \exp[-\beta x], \tag{8.74}$$

$$\psi_{III} = F \exp[i\alpha x]. \tag{8.75}$$

Note that ψ_I and ψ_{III} each contain terms to allow for particle movement in *either* direction in regions I and, II.

The problem now is to evaluate the constants *A, B, C, D,* and *F* in ψ_I, ψ_{II}, and ψ_{III} through the application of appropriate boundary conditions. In order for the entire function to be well- behaved, that is, continuous and smooth, it is necessary that ψ and also $d\psi/dx$ be continuous at the position $x = 0$ and $x = a$. The *continuity conditions* are then

$$\psi_I = \psi_{II} \text{ (at } x = 0), \tag{8.76}$$

$$\frac{d\psi_I}{dx} = \frac{d\psi_{II}}{dx} \text{ (at } x = 0), \tag{8.77}$$

$$\psi_{II} = \psi_{III} \text{ (at } x = a), \tag{8.78}$$

$$\frac{d\psi_{II}}{dx} = \frac{d\psi_{III}}{dx} \text{ (at } x = a). \tag{8.79}$$

Substitution of ψ_I, ψ_{II}, and ψ_{III} from Eqs. (8.73), (8.74) and (8.75) into the four continuity equations yields, respectively,

$$A + B = C + D, \tag{8.80}$$

$$i\alpha(A - B) = \beta(C - D), \tag{8.81}$$

$$C \exp[\beta a] = F \exp[i\alpha a], \tag{8.82}$$

$$\beta C \exp[\beta a] - \beta D \exp[-\beta a] = i\alpha F \exp[i\alpha a]. \tag{8.83}$$

The above set of four equations contains five unknowns, so that we cannot directly solve for each of the unknowns. We may, however, solve for any one of the unknowns in terms of any other. For our purpose, it is most instructive to solve for A in terms of F, so that we may then derive an expression for the probability that the particle penetrates, or is transmitted through, the barrier. The *transmission coefficient* T is defined as

$$\tau = \frac{|F|^2}{|A|^2}. \tag{8.84}$$

Since $|A|^2$ is the probability density for the particle moving from left to right in region I so that it impinges on the left side of the barrier, and $|F|^2$ is the probability density for the particle moving to the right in region III *after emergence* from the barrier, the ratio $|F|^2/|A|^2$ is the probability that a single particle penetrates or is transmitted *through* the barrier.

For a system of many particles, the transmission coefficient τ is the *fraction* of particles which penetrates the barrier and emergence in region III.

Solution of Eqs. (8.80) and (8.81) for A and B in terms of C and D yields

$$A = \frac{C(i\alpha + \beta)}{2i\alpha} + \frac{D(i\alpha - \beta)}{2i\alpha}, \tag{8.85}$$

$$B = \frac{C(i\alpha - \beta)}{2i\alpha} + \frac{D(i\alpha + \beta)}{2i\alpha}, \tag{8.86}$$

and solution of Eqs. (8.82) and (8.83) for C and D in terms of F gives

$$C = \frac{F(\beta + i\alpha)}{2\beta}[\exp(i\alpha - \beta)\alpha], \tag{8.87}$$

$$D = \frac{F(\beta - i\alpha)}{2\beta}[\exp(i\alpha + \beta)\alpha], \tag{8.88}$$

Substituting Eqs. (8.87) and (8.88) into Eq. (8.85)

$$A = F\left[\frac{(\beta^2 + 2i\alpha\beta - \alpha^2)}{4i\alpha\beta}[\exp(i\alpha - \beta)\alpha] + \frac{\alpha^2 + 2i\alpha\beta - \beta^2)}{4i\alpha\beta}[\exp(i\alpha - \beta)\alpha]\right]$$

$$= \frac{Fe^{i\alpha a}}{4i\alpha\beta}[(\beta^2 + 2i\alpha\beta - \alpha^2)\exp(-\beta\alpha) + (\alpha^2 + 2i\alpha\beta - \beta^2)\exp(\beta\alpha)]$$

$$= \frac{Fe^{i\alpha\beta}}{4i\alpha\beta}\,[2i\alpha\beta(\exp(\beta\alpha) + \exp(\beta\alpha) + (\alpha^2 - \beta^2)(\exp(\beta\alpha) - \exp(-\beta\alpha)] \quad .(8.89)$$

Multiplication of each side of Eq. (8.89) by its complex conjugate yields

$$|A|^2 = |F|^2\left[\frac{\exp(\beta\alpha + \exp(-\beta\alpha)}{4} + \frac{(\alpha^2 - \beta^2)^2}{16\alpha^2\beta^2}\exp(\beta\alpha) - \exp(-\beta\alpha)^2\right] \quad (8.90)$$

Examination of Eq. (8.90) reveals that Γ, which is given by $|F|^2/|A|^2$, ***goes to zero only when*** $\beta\alpha$ ***is infinite.*** Eq. (8.90) may be simplified as follows :

1. If α and β are of similar magnitude, $(\alpha^2 - \beta^2)$ is small, so that the second term in Eq. (8.90) may be neglected relative to the first term.
2. If $\beta\alpha > 1$, $e^{-\beta a}$ is small compared to $e^{\beta a}$, so that the first term in the brackets becomes $e^{2\beta a}/4$.

Thus Eq. (8.90) may be written in the *approximate* form

$$\Gamma = \frac{|F|^2}{|A|^2} \cong 4e^{-2\beta a}.$$

Substitution of β from Eq. (8.72) gives

$$\Gamma = 4\exp\left[\frac{-4\pi a}{h}\sqrt{2m(V_0 - E)}\right]. \quad (8.91)$$

It is clear that unless V_0, a, or m is infinite, Γ will not be equal to zero and, therefore, *the particle for any value of E will have a finite probability for penetrating the barrier.* Note also that Γ increases as m decreases, or as the thickness a or height V_0 of the barrier decreases.

Experimental evidence indicates that barrier penetration, referred to as *quantum-mechanical tunneling,* actually does occur.

The most direct evidence is found in the radioactive disintegration of atomic nuclei by alpha decay.

Because of the small mass of the electron and the exponential dependence of Γ on m, tunneling appears to be important in many processes in which electrons are transferred, such as electrode reactions, oxidation-reduction reactions in solution, and cold emission of electrons.

The umbrella-like *inversion* of pyramidal molecules such as NH_3, PH_3 and AsH_3 is believed to occur through a process in which the N atom, tunnels through the potential energy barrier associated with the plane of the three H atoms. Experimental spectral observations indicate that the energy of inversion is not sufficient to allow the N atoms to pass *over* the barrier.

SECTION—III

ATOMIC SYSTEMS

9

The Hydrogen Atom

THE SCHRODINGER EQUATION FOR THE HYDROGEN ATOM

We shall now deal with the application of Quantum mechanical methods, to systems which consist of a positively charged nucleus and in electron moving around it as is found in the hydrogen atom as well as in ions like He^+, Li^{2+}, Be^{3+} etc. the Schrodinger equation for the hydrogen-like atom is

$$\frac{-h^2}{8\pi^2\mu}\left(\frac{\partial^2\psi}{\partial x^2}+\frac{\partial^2\psi}{\partial y^2}+\frac{\partial^2\psi}{\partial z^2}\right)-\frac{Ze^2\psi}{r}=E\psi. \qquad (9.1)$$

where μ is the reduced mass of the atom.

The equation is solved as described in the following text.

Separation of Variables in the Equation of Relative Motion

It is convenient to transform Eq. (9.1) into an equation involving spherical polar coordinates (r, θ, ϕ) rather than rectangular coordinates. The relationship between the two coordinate systems is shown in Fig. 9.1.

Transformation equations are

$$x = r\sin\theta\cos\phi$$

$$y = r\sin\theta\sin\phi$$

$$z = r\cos\theta$$

Appropriate substitution of the transformation equations into Eq. (9.1) leads to the spherical polar form of the Schrodinger equation :

$$\frac{-h^2}{8\pi^2\mu r^2}\left[\frac{\partial}{\partial r}\left(r^2\frac{\partial\psi}{\partial r}\right)+\frac{1}{\sin\theta}\frac{\partial}{\partial\theta}\left(\sin\theta\frac{\partial\psi}{\partial\theta}\right)+\frac{1}{\sin^2\theta}\frac{\partial^2\psi}{\partial\varphi^2}\right]-\frac{Ze^2}{r}\psi=E\psi \quad ,(9.2)$$

where ψ is now a function of r, θ, and ϕ.

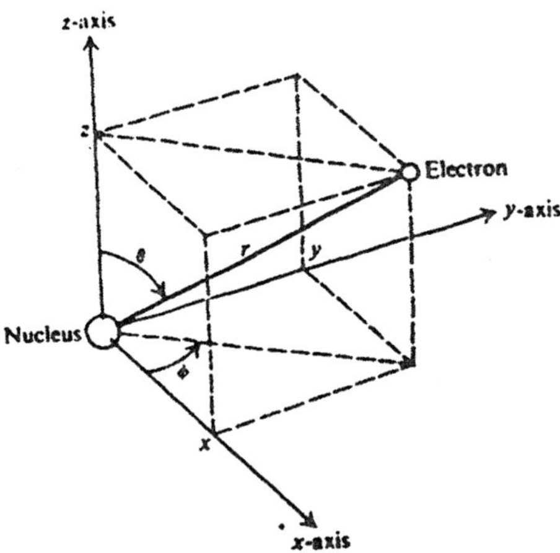

Fig. 9.1. The relationship among the polar coordinates (r, θ, ϕ) and the cartesian coordinates (x, y, z) of the electron relative to the nucleus.

In order to solve Eq. (9.2) we will separate the three- dimensional Schrodinger equation into three independent and soluble one-dimensional eigenvalue equations by assuming that the wave function $\psi(x, y, z)$ could be expressed as the product of three wave functions, each dependent on only one of the coordinates. That is assume that $\psi(r, \theta, \phi)$ may be expressed as the product of three independent functions, each dependent on only one of the coordinate positions,

$$\psi(r, \theta, \phi) = R(r)\Theta(\theta)\Phi(\varphi). \quad (9.3)$$

The function $R(r)$ depends *only* on r, the function $\Theta(\theta)$ depends *only* on θ, and the function $\Phi(\phi)$ depends *only* on ϕ. Incorporate Eq. (9.3) into Eq. (9.2) and determine whether or not a suitable separation results. Substitution of Eq. (9.3) with simplified notation into Eq. (9.2) gives

$$\frac{-h^2}{8\pi^2\mu^2}\left[\Theta\Phi\frac{d}{dr}\left(r^2\frac{dR}{dr}\right)+R\Theta\frac{1}{\sin\theta}\frac{d}{d\theta}\left(\sin\theta\frac{d\Theta}{d\theta}\right)+R\Theta\frac{1}{\sin^2\theta}\frac{d^2\Phi}{d\varphi^2}\right]-\frac{Ze^2}{r}R\Theta\Phi$$

$$= ER\Theta\Phi. \tag{9.4}$$

If we multiply the entire equation by $\sin^2 0/R\Theta\Phi$ and rearrange, Eq. (9.4) yields :

$$\frac{\sin^2\theta}{R}\frac{d}{dr}\left(r^2\frac{dR}{dr}\right)+\frac{\sin\theta}{\Theta}\frac{d}{d\theta}\left(\sin\theta\frac{d\Theta}{d\theta}\right)$$
$$+\frac{1}{\Phi}\frac{d^2\Phi}{d\varphi^2}+\frac{8\pi^2\mu^2\sin^2\theta}{h^2}\left(E+\frac{Ze^2}{r}\right)=0 \tag{9.5}$$

But note that the third term of Eq. (9.5) is the only term involving ϕ. Since this term is independent of variations in r and θ, and the other terms are independent of variations in θ, the term

$$\frac{1}{\Phi}\cdot\frac{d^2\Phi}{d\varphi^2}$$

must be a constant, and the sum of the other terms in the equation must be the negative of this same constant. Calling the constant A, we may write

$$\frac{d^2\Phi}{d\varphi^2}=A\Phi. \tag{9.6}$$

Equation (9.6) **is an eigenvalue equation whose satisfactory solution will yield Φ as eigenfunctions and A as eigenvalues.** Substitution of the constant A back into Eq. (9.5) followed by division by $\sin^2\theta$ yields

$$\frac{1}{R}\cdot\frac{d}{dr}\left(r^2\frac{dR}{dr}\right)+\frac{1}{\Theta\sin\theta}\cdot\frac{d}{d\theta}\left(\sin\theta\frac{d\Theta}{d\theta}\right)+\frac{A}{\sin^2\theta}+\frac{8\pi^2\mu r^2}{h^2}\left(E+\frac{Ze^2}{r}\right)=0. \tag{9.7}$$

But the first and fourth terms of Eq. (9.7) depends only on r and are independent of θ, and the second and third terms depend only on θ and are independent of r. It follows therefore, that, since the sum of all four terms is a constant (zero), the sum of the set of terms involving r and the sum of the set of terms involving θ must each independently be equal to constants. Furthermore, the value of the constant for the r terms must be the negative of the value of the constant for the θ terms, since the sum of the two constants must be equal to zero. Thus, arbitrarily letting the constant for the θ terms equal $(-B)$, we may write

$$\frac{1}{\Theta \sin\theta}\frac{d}{d\theta}\left(\sin\theta \frac{d\Theta}{d\theta}\right) + \frac{A}{\sin^2\theta} = -B \tag{9.8}$$

or

$$\frac{1}{\sin\theta}\frac{d}{d\theta}\left(\sin\theta \frac{d\Theta}{d\theta}\right) + \frac{A\Theta}{\sin^2\theta} = -B\Theta. \tag{9.9}$$

Equation (9.9) **is an eigenvalue equation whose solution will yield Θ as eigenfunctions and $-B$ as eigenvalues.**

In addition, since the sum of the terms involving r Eq. (9.9) must be equal to B, we may write

$$\frac{1}{R}\frac{d}{dr}\left(r^2 \frac{dR}{dr}\right) + \frac{8\pi^2\mu r^2}{h^2}\left(E + \frac{Ze^2}{r}\right) = B. \tag{9.10}$$

Multiplication by $Rh^2/8\pi^2\mu r^2$ gives

$$\frac{h^2}{8\pi^2\mu r^2}\frac{d}{dr}\left(r^2 \frac{dR}{dr}\right) + ER + \left(\frac{Ze^2}{r}\right)R = \left(\frac{Bh^2}{8\pi^2\mu r^2}\right)R, \tag{9.11}$$

and rearrangement of Eq. (9.11) yields,

$$\frac{h^2}{8\pi^2\mu r^2}\frac{d}{dr}\left(r^2 \frac{dR}{dr}\right) + \left(\frac{Ze^2}{r}\right)R - \left(\frac{Bh^2}{8\pi^2\mu r^2}\right)R = -ER. \tag{9.12}$$

Eq. (9.12) **is an eigenvalue equation whose satisfactory solution yields R as eigenfunctions and $-R$ as eigenvalues.**

Thus we are able to separate the Schrodinger equation involving r, θ, and ϕ (Eq. 9.2) into three separate eigenvalue equations, each involving only one of the coordinates, has been successful. In fact, *such a separation is always successful when V is a function of r only.* Solution of Eq. (9.6) gives the allowed forms for the function $\Phi(\phi)$, solution of Eq. (9.9) gives the allowed forms for the function $\Theta(\theta)$, and solution of Eq. (9.12) gives the allowed forms for the function $R(r)$.

SOLUTION OF THE Φ EQUATION

The form of the Φ equation is

$$\frac{d^2\Phi}{d\varphi^2} = A\Phi, \tag{9.6}$$

An acceptable eigenfunction, which may be verified easily by substitution, is

$$\Phi(\varphi) = \alpha \exp (A^{1/2}\varphi), \tag{9.13}$$

where α is a constant. In order that Φ be continuous and single-valued, it is necessary that Φ have the same value at ϕ as at $(\phi + 2\pi)$ since these are identical angular positions. Otherwise, two different values of Φ at the same position would lead to two different values of $\psi = \Phi\Theta R$ and hence to two different values of $\psi^*\psi$, the probability density, at the same position, which is not physically acceptable. Let us then set Φ at $\phi = 0$ equal to Φ at $\phi = 2\pi$:

$$\Phi(0) = \Phi(2\pi),$$

or

$$\alpha\exp (A^{1/2}(0) = \alpha\exp[A^{1/2} (2\pi)], \tag{9.14}$$

which apparently cannot be so if A is a real number, since $2\pi \neq 0$. Equation (9.14) may be true only if Φ is a periodic function involving an *imaginary* exponent, that is, if $A^{1/2}$ is imaginary. Therefore, let's arbitrarily write

$$A^{1/2} = im, \text{ or } A = -m^2,$$

where m is a constant. Substitution for $A^{1/2}$ in Eq. (9.14) yields

$$\exp [im(0)] = \exp [im(2\pi)],$$

from which we obtain

$$1 = \cos (m2\pi) + i \sin (m2\pi),$$

this requires that

$$m = 0, \pm 1, \pm 2, \pm 3, ..., \pm\infty. \tag{9.15}$$

There are therefore, an infinite number of discrete solutions to the Φ-equation, each of the form

$$\Phi(phi) = \exp (im\varphi), \tag{9.16}$$

where m is restricted to integers as given in Eq. (9.15).

The total wave function $\psi(\phi, \theta, r)$ is normalized if we normalize $\Phi(\phi)$, $\Theta(\theta)$, and $R(r)$ individually. For normalized Φ function, the constant α may be determined easily from the condition

$$\int_0^{2\pi} \Phi^*(\varphi)\Theta(\varphi)\, d\varphi = 1.$$

Substitution of Eq. (9.16) into the above equation yields

$$\alpha^2 \int_0^{2\pi} e^{-im\varphi}\, e^{im\varphi}\, d\varphi = \alpha_2^{2\pi}\, d\varphi = \alpha^2 2\pi = 1,$$

from which

$$\alpha = 1/\sqrt{2\varphi},$$

so that Eq. (9.16) may be written, for *normalized* Φ functions, as

$$\Phi_m(\varphi) = \frac{1}{\sqrt{2\pi}} \exp(im\varphi). \tag{9.17}$$

It is clear that the form of the Φ function depends on the value of the integer m. In effect, m serves to quantize the eigenvalues A in Eq. (9.6) such that $A = -m^2$. The integer m is called the *magnetic quantum number.*

SOLUTION OF THE Φ EQUATION

Let us begin by substituting $-m^2$ for A in Eq. (9.9) and rearranging, which gives

$$\frac{1}{\sin\theta}\frac{d}{d\theta}\left(\sin\theta\frac{d\Theta}{d\theta}\right) - \frac{m^2\Theta}{\sin^2\theta} + B\Theta = 0 \tag{9.18}$$

It is convenient to define the new independent variable z (which is not the cartesian coordinate) as

$$z = \cos\theta$$

so that

$$1 - z^2 = 1 - \cos^2\theta = \sin^2\theta. \tag{9.19}$$

Lets also replace $\Theta(\theta)$ with an equivalent function of z, $P(z)$, defined so that

$$P(z) = \Theta(\theta). \tag{9.20}$$

Then

$$\frac{d\Theta}{d\theta} = \frac{dP}{d\theta} = \frac{dP}{dz}\cdot\frac{dz}{d\theta} = \frac{dP}{dz}\frac{d(\cos\theta)}{d\theta} = -\sin\theta\frac{dP}{dz}, \tag{9.21}$$

from which, since $\Theta = P$, we may write the operator

$$\frac{d}{d\theta} = -\sin\theta \frac{d}{dz}. \tag{9.22}$$

Substitution of Eqs. (9.20), (9.21), and (9.22) into Eq. (9.18) yields

$$\frac{1}{\sin}\theta\left(-\sin\theta\frac{d}{dz}\right)\left[\sin\theta\left(-\sin\theta\frac{dP}{dz}\right)\right] - \frac{m^2P}{\sin^2\theta} + BP = 0,$$

or

$$\frac{d}{dz}\left(\sin^2\theta\frac{dP}{dz}\right) + \left(B - \frac{m^2}{\sin^2\theta}\right)P = 0. \tag{9.23}$$

Further substitution of Eq. (9.19) into Eq. (9.23) gives

$$\frac{d}{dz}\left[(1 - z^2)\frac{dP}{dz}\right] + \left(B - \frac{m^2}{1 - z^2}\right)P = 0,$$

Differentiation of the term in brackets yields

$$(1 - z^2)\frac{d^2P}{dz^2} - 2z\frac{dP}{dz} + \left(\frac{B - m^2}{1 - z^2}\right)P = 0. \tag{9.24}$$

If we now restrict the constant B to values given by

$$B = l(l + 1)$$

where l is an integer, we may write Eq. (9.24) follows :

$$(1 - z^2)\frac{d^2P}{dz^2} - 2z\frac{dP}{dz} + \left[l(l + 1) - \frac{m^2}{1 - z^2}\right]P = 0, \tag{9.25}$$

which has the form of a well-known differential equation called the *associated Legendre equation,* whose solutions, P, are given as the *associated Legendre polynomials* $P_l^{|m|}$ of *degree* l and *order* $|m|$, as follows :

$$P = P_l^{|m|}(z) = (1 - z^2)^{|m|/2}\frac{d^{|m|}}{dz^{|m|}}P_l(z). \tag{9.26}$$

l and $|m|$ in $P_l^{|m|}(z)$ are meant to indicate that the forms of the associated Legendre polynomials, which are functions of z, depend on the value of the positive integer l and on the absolute value $|m|$ of the integer m, which we have already defined i the previous section. Note also that $P_l^{|m|}(z)$ is defined in terms of *another* function $P_l(z)$, which is called the *Legendre polynomial.* The Legendre polynomials are also functions of z, whose forms depend on the value of the positive integer l and are given by

$$P_l(z) = \frac{1}{2^l(l!)dz^l}(z^2 - 1)^l. \tag{9.27}$$

The conditions of the solution require that l, which is called *azimuthal quantum number,* be restricted to the integers :

$$l = 0, 1, 2, 3, ...,$$

and that m, the magnetic quantum number be restricted to the integers:

$$m = 0, \pm 1, \pm 2, ..., \pm l,$$

or

$$|m| = 0, 1, 2, ..., l.$$

Note that even though the value of the integer m was not restricted to any maximum value in the previous solution of the Φ equation, *the present solution of the Φ equation requires that $|m| \leq l$ if the functions are not to vanish.* Some of the Legendre polynomials $P_l(z)$ for selected values of l are

$$l = 0, \quad P_0(z) = 1,$$

$$l = 1, \quad P_1(z) = z,$$

$$l = 2, \quad P_2(z) = \tfrac{3}{2}z^2 - \tfrac{1}{2},$$

$$l = 3, \quad P_3(z) = \tfrac{5}{2}z^3 - \tfrac{3}{2}z.$$

Thus, if we require that $B = l(l + 1)$, we are able to manipulate the Θ equation into the form of the associated Legendre equation. It then follows that the $\Theta(\theta)$ eigenfunctions are identical to the solutions of the associated Legendre equation, that is, identical to the associated Legendre polynomials as given by Eq. (9.26). However, in order to normalize the functions, we must further introduce the constant β such that

$$\Theta(\theta)_{\text{normalized}} = \beta P_l^{|m|}(z). \tag{9.28}$$

The value of β is then determined from the normalization condition,

$$\beta^2 \int_{-1}^{+1} P_l^{|m|} P_l^{|m|} \, dz = 1,$$

where the variables $z = \cos\theta$, is confined between the limits $z = -1$ and $z = +1$. By appropriate substitution and integration it mav be shown that

$$\beta = \left[\frac{(2l+1)}{2} \frac{(l-|m|)!}{(l+|m|)!} \right]^{1/2}, \tag{9.29}$$

so that the normalized Θ functions may be written, by substitution of Eq. (9.29) into Eq. (9.30), as

$$\Theta_{lm}(\theta) = \left[\frac{(2l+1)}{2} \frac{(l-|m|)!}{(l+|m|)!} \right]^{1/2} P_l^{|m|}(z). \tag{9.30}$$

We have now added the subscripts l and m to Θ to indicate that the Θ function depends on the values of the quantum numbers l and m. Through substitution of $P_l(z)$ from Eq. (9.27) into Eq. (9.26), followed by substitution of $P_l^{|m|}(z)$ from the resultant equation into Eq. (9.30), we obtain the final explict normalized Θ function as

$$\Theta_{lm}(\theta) = \frac{(-1)^l}{2^l(l!)} \left[\frac{(2l+1)}{2} \frac{(l-|m|)!}{(l+|m|)!} \right]^{1/2} \sin^{|m|}\theta \frac{d^{l+|m|}(\sin^{2l}\theta)}{(d\cos\theta)^{l+|m|}}. \tag{9.31}$$

Although Eq. (9.31) seems to be rather formidable, the Θ functions reduce to relatively simple forms on substitution of specific values of the quantum numbers l and m.

Solution of the R Equation

We will begin by substitution $l(l+1)$ for B in Eq. (9.12) and then rearranging to

$$\frac{1}{r^2}\frac{d}{dr}\left(r^2 \frac{dR}{dr}\right) - \frac{l(l+1)R}{r^2} + \frac{8\pi^2\mu ER}{h^2} + \frac{8\pi^2\mu Ze^2R}{h^2 r} = 0. \tag{9.32}$$

We know that E is constant and *negative* for the *bound* electron; we now define two new constants α and λ, related to the energy E, such that

$$\alpha^2 = -8\pi^2\mu E/h^2, \tag{9.33}$$

and

$$\lambda = 4\pi^2\mu Ze^2/h^2\alpha. \qquad (9.34)$$

Substitution of Eqs. (9.33) and (9.34) into Eq. (9.32) yields

$$\frac{1}{r^2}\frac{d}{dr}\left(r^2\frac{dR}{dr}\right) - \frac{l(l+1)R}{r^2} - \alpha^2 R + \frac{2\alpha\lambda R}{r} = 0. \qquad (9.35)$$

In order to change the equation to a more familiar form, we find it convenient now to introduce the new independent variable ρ where

$$\rho = 2\alpha r, \qquad (9.36)$$

and the new function ρ, $S(\rho)$, where

$$S(\rho) = R(r), \qquad (9.37)$$

which are to be substituted into Eq. (9.35). In order to transform the differential terms, we may write

$$\frac{dR(r)}{dr} = \frac{dS(\rho)}{dr} = \frac{dS(\rho)}{d\rho}\frac{d\rho}{dr} = 2\alpha\frac{dS(\rho)}{d\rho}, \qquad (9.38)$$

and from Eq. (9.38), since $R(r) = S(\rho)$,

$$\frac{d}{dr} = 2\alpha\frac{d}{d\rho}. \qquad (9.39)$$

Substitution of Eqs. (9.36), (9.37), (9.38), and (9.39) into Eq. (9.35) gives

$$\left(\frac{4\alpha^2}{\rho^2}\right)2\alpha\frac{d}{d\rho}\left[\left(\frac{\rho^2}{4\alpha^2}\right)2\alpha\frac{dS(\rho)}{d\rho}\right] - \frac{4\alpha^2}{\rho^2}l(l+1)S(\rho) - \alpha^2 S(\rho),$$
$$+\left(\frac{2\alpha}{\rho}\right)2\alpha\lambda S(\rho) = 0$$

and division by $4\alpha^2$ gives

$$\frac{1}{\rho^2}\frac{d}{d\rho}\left[\rho^2\frac{dS(\rho)}{d\rho}\right] + \left[\frac{-l(l+1)}{\rho^2} - \frac{1}{4} + \frac{\lambda}{\rho}\right]S(\rho) = 0, \qquad (9.40)$$

where ρ, which is directly proportional to r, may assume positive values between zero and infinity. completion of the differential indicated in Eq. (9.40) yields

$$\frac{d^2S(\rho)}{d\rho^2} + \frac{2}{\rho}\frac{dS(\rho)}{d\rho} + \left[\frac{-l(l+1)}{\rho^2} - \frac{1}{4} + \frac{\lambda}{\rho}\right]S(\rho) = 0. \qquad (9.41)$$

In order to find a solution for Eq. (9.41) let's first look at the form of the equation as ρ approaches infinity. As ρ– inf, Eq. (9.41) becomes

$$\frac{d^2S(\rho)_\infty}{d\rho^2} = \frac{S(\rho)_\infty}{4}, \tag{9.42}$$

which is an *asymptotic equation* which may be used to obtain the *asymptotic solution* $S(\rho)\infty$ The solution of Eq. (9.42) may be easily verified by substitution to be

$$S(\rho)_\infty = c \exp(\pm\rho/2), \tag{9.43}$$

where c is an arbitrary constant in the asymptotic solution, but may otherwise *generally* be a function of ρ, $F(\rho)$, such that $F(\rho) \rightarrow c$ as $\rho \rightarrow \infty$. That is, in general

$$S(\rho) = \exp(\pm\rho/2)F(\rho)$$

Positive or negative exponent both leads to the same final solution so that we may have

$$S(\rho) = \exp(-\rho/2)F(\rho) \tag{9.44}$$

where $F(\rho)$ is presumably a simpler function than $S(\rho)$. Now let's transform Eq. (9.41) into an equation involving $F(\rho)$ rather than $S(\rho)$. Differentiating Eq. (9.44), we obtain

$$\frac{dS(\rho)}{d\rho} = \exp(-\rho/2)\frac{dF(\rho)}{d\rho} - ½\exp(-\rho/2)F(\rho), \tag{9.45}$$

and further differentiation and collection of terms yields

$$\frac{d^2S(\rho)}{d\rho^2} \exp(-\rho/2)\left[\frac{d^2F(\rho)}{d\rho^2} - \frac{dF(\rho)}{d\rho} + \frac{F(\rho)}{4}\right]. \tag{9.46}$$

Substitution of Eqs. (9.44), (9.45), and (9.46) into Eq. (9.41) yields

$$\exp(-\rho/2)\left[\frac{d^2F(\rho)}{d\rho^2} - \frac{dF(\rho)}{d\rho} + \frac{F(\rho)}{4}\right] + \frac{2}{\rho}\exp(-\rho/2)\left[\frac{dF(\rho)}{d\rho} - \frac{F(rho)}{2}\right]$$
$$+ \left[\frac{-l(l+1)}{\rho^2} - \frac{1}{4} + \frac{\lambda}{\rho}\right]\exp(-\rho/2)F(\rho) = 0,$$

and division by $\exp(-\rho/2)$ followed by rearrangement leads to

$$\frac{d^2F(\rho)}{d\rho^2} + \left(\frac{2}{\rho} - 1\right)\frac{df(\rho)}{d\rho} + \left[\frac{\lambda}{\rho} - \frac{l(l+1)}{\rho^2} - \frac{1}{\rho}\right]F(\rho) = 0, \quad (9.47)$$

where $F(\rho)$ may vary from zero to infinity.

In order to examine the solution to Eq. (9.47) as $\rho \to 0$, we now define still another new function $L(\rho)$, such that

$$F(\rho) = \rho^3 L(\rho), \quad (9.48)$$

where the constant s is to be determined from the nature of the equation at $\rho = 0$. Differentiation of Eq. (9.48) yields

$$\frac{dF(\rho)}{d\rho} = \rho^3 \frac{dL(\rho)}{d\rho} + s\rho^{s-1}L(\rho), \quad (9.49)$$

and further differentiation gives

$$\frac{d^2F(\rho)}{d\rho^2} = \rho^3 \frac{d^2L(\rho)}{d\rho^2} + 2s\rho^{s-1}\frac{dL(\rho)}{d\rho} + s(s-1)\rho^{s-2}L(\rho). \quad (9.50)$$

Substitution of Eqs. (9.48), (9.49), and (9.50) into Eq. (9.47), followed by collection of terms, yields

$$\rho^s \frac{d^2L(\rho)}{d\rho^2} + (2s\rho^{s-1} + 2\rho^{s-1} - \rho^s)\frac{dL(\rho)}{d\rho}$$
$$+ [s(s-1)\rho^{s-2} + 2s\rho s - 2 - s\rho^{s-1} + \lambda\rho^{s-1}$$
$$- l(l+1)\rho^{s-2} - \rho^{s-1}]L(\rho) = 0.$$

Division by ρ^{s-2} with further collection of terms yields

$$\rho^2 \frac{d^2L(\rho)}{d\rho^2} + [2s + 2 - \rho)\rho]\frac{dL(\rho)}{d\rho}$$
$$+ [s(s+1) - l(l+1) + (\lambda - s -. 1)\rho]L(\rho) = 0 \quad (9.51)$$

Note that Eq. (9.51) must be satisfied by *all* values of ρ, including $\rho = 0$. In order to determine the value of the constant s let's then substitute $\rho = 0$ into Eq. (9.51), which yields

$$s(s+1) - l(l+1) = 0,$$

which, in turn, is satisfied by the two roots

$$s = l \text{ or } s = -(l+1).$$

But the negative solution, $s = -(l + 1)$, must be rejected since, according to Eq. (9.48), a negative value of s would cause $F(\rho)$ to approach infinity (to diverge) as ρ approaches zero and would consequently cause divergence of $R(r)$ and $\psi(\varphi, \theta, r)$ as $\rho \to 0$. T'en, accepting the solution $s = l$, we may write Eq. (9.48) as

$$F(\rho) = \rho^l L(\rho). \tag{9.52}$$

Let's now substitute l for s in Eq. (9.51)

$$\frac{\rho^2 d^2 L(\rho)}{d\rho^2} + [(2l + 2 - \rho)\rho]\frac{dL(\rho)}{d\rho}$$
$$+ [l(l + 1) - l(l + 1) + (\lambda - l - 1)\rho]L(\rho) = 0$$

Division of ρ, followed by collection of terms, yields the final differential equation

$$\frac{\rho d^2 L(\rho)}{d\rho^2} + [2(l + 1) - \rho]\frac{dL(\rho)}{d\rho} + (\lambda - l - 1)L(\rho) = 0. \tag{9.53}$$

We may now write $R(r)$ as

$$R(r) = S(\rho) = \exp(-\rho/2)F(\rho) = \rho^l \exp(-\rho/2)L(\rho) \tag{9.54}$$

Eq. (9.12) is now in form which is similar to that of a well- known equation, called the *associated equation* which is

$$\frac{\rho d^2 L_k^j(\rho)}{d\rho^2} + (j + 1 - \rho)\frac{dL_k^j(\rho)}{d\rho} + (k - j)L_k^j(\rho) = 0. \tag{9.55}$$

The solutions of Eq. (9.55) are the *associated Laguerre polynomials,* of degree $k - j$ and order j :

$$L_k^j(\rho) = \frac{d^j}{d\rho^j} L_k(\rho), \tag{9.56}$$

where k and j are positive integers, and where $L_k(\rho)$, the *Lagurre polynomials* of degree k, are given by

$$L_k(\rho) = e^\rho \frac{d^k}{d\rho^k} \rho^k e^{-\rho}. \tag{9.57}$$

Some of the Laguerre polynomials obtained through substitution of selected k values in Eq. (9.57) are

$$k = 0, \quad L_0(\rho) = 1,$$

$$k = 1, \quad L_1(\rho) = 1 - \rho,$$

$$k = 2, \quad L_2(\rho) = 2 - r\rho + \rho^2,$$

$$k = 3, \quad L_3(\rho) = 6 - 18\rho = 9\rho^2 - \rho^3, \qquad (9.58)$$

Comparison of Eq. (9.53) with the associated Laguerre equation, Eq. (9.55), indicates that the solutions $L(\rho)$ for Eq. (9.53) are given by the associated Laguerre polynomials $L_k^j(\rho)$ *if*

$$2(l + 1) - \rho = j + 1 - \rho,$$

and

$$\lambda - l - 1 = k - j,$$

from which

$$j = 2l + 1, \qquad (9.59)$$

and

$$k = \lambda = l. \qquad (9.60)$$

Since the solution of the associated Laguerre equation demands that k be a positive integer, it follows from Eq. (9.60) that λ *must also be a positive integer,* which we will now define as n, the *principal quantum number.* Then, from Eq. (9.60),

$$k = n + l \qquad (9.61)$$

and the associated Laguerre polynomials which satisfy Eq. (9.53) written, by substitution of Eqs. (9.55) and (9.61) into Eq. (9.56) as

$$L(\rho) = L_k^j(\rho) = L_{n+l}^{2l+1}(\rho). \qquad (9.62)$$

Substitution of Eq. (9.62) into Eq. (9.54) yields, for *normalized* $R(r)$ functions,

$$R(r) = \gamma\rho^l e^{-\rho/2} L_{n+l}^{2l+1}(\rho), \qquad (9.63)$$

where γ is introduced as a noramalization constant and where $\rho = 2\alpha r$. Note that from Eq. (9.34), where $\lambda = n$, we may write

$$\alpha = 4\pi^2\mu Ze^2/nh^2 = Z/na_0,$$

where

$$a_0 = h^2/4\pi^2\mu e^2 \qquad (9.64)$$

The term a_0 has been defined previously as the radius of the first Bohr orbit in the hydrogen atom. Then, we may write the variable ρ as

$$\rho = 2\alpha r = (2Z/na_0)r. \qquad (9.65)$$

Now let's examine the allowed values for the positive integer n which we have called the principal quantum number. If the associated Laguerre polynomial is not to vanish, it is necesssary, according to Eq. (9.56), that

$$j \le k$$

which on substitution of $j = 2l + 1$ and $k = n + l$ becomes

$$2l + 1 \le n + l,$$

so that

$$l + 1 \le n \text{ or } l \le n - 1.$$

Since l is allowed the values 0, 1, 2, 3, ..., the principal quantum number n obviously may not have a value of zero but it restricted to the positive integers given by

$$n = 1, 2, 3, \ldots, \infty,$$

and since the value of l may not exceed $n - 1$, l is restricted to the values

$$l = 1, 2, \ldots, (n - 1).$$

As the final step in determining a completely satisfactory R equation, we must normalize the expression give by Eq. (9.63) between the limits $\rho = 0$ and $\rho = \infty$. It may be shown that such normalization leads to the value for the normalization constant :

$$\gamma = -\left\{\left(\frac{2Z}{na_0}\right)^3 \frac{n - l - 1)!}{2n[n + l)]^3}\right\}^{1/2},$$

where the (–) sign is used to make the functions positive, so that the final normalized radial functions are

$$R_{nl}(r) = -\left\{\left(\frac{2Z}{na_0}\right)^3 \frac{n - l - 1)!}{2n[n + l)]^3}\right\}^{1/2} \rho^l e^{-\rho/2} L_{n+l}^{2l+1}(\rho), \qquad (9.66)$$

where

$$\rho = (2Z/na_0)r.$$

THE PRINCIPAL QUANTUM NUMBER n

We will now examine the dependence of the energy E of the atom on the principal quantum number n. If we substitute n for λ in Eq. (9.34), we obtain

$$\alpha = 4\pi^2\mu Ze^2/nh^2. \qquad (9.67)$$

Substitution of α from Eq. (9.67) into Eq. (9.33), followed by rearrangement, yields

$$E = -2\pi^2\mu Z^2e^4/n^2h^2, \qquad (9.68)$$

which is in *exact agreement* with the Bohr equation. According to Eq. (9.68), the energy of relative motion of the hydrogen atom or hydrogen-like ion is quantized and depends on the value of the principal quantum number n, but is independent of the values of the other two quantum number l and m. Since the principal quantum number n appears *only* in the radial wave function $R(r)$, and does not appear in $\Theta(\theta)$ nor in $\Phi(\phi)$, we can say that the energy of relative motion of the atom is related to the distance of the electron from the nucleus and does not depend on any particular angular orientation.

The atom may have its lowest energy in quantum states for which $n = 1$. But when $n = 1$, l may have *only* the value zero, and m may also have *only* a value of zero. We designate this first allowed quantum state as the (100) state. In general, the form (nlm) is used to designate quantum states. The energy in the lowest quantum state is

$$e_{(nlm)} = E_{(100)} = \frac{-2\pi^2\mu e^4}{h^2}$$

$$= \frac{-2\pi^2(9.104\times 10^{-31}\,\text{kg})(1.66\times 10^{-19}\text{C})^4}{(6.626\times 10^{-34}\text{Js})^2}$$

$$= -2.178\times 10^{-18}\text{J} \qquad (9.69)$$

The atom has its next highest energy when $n = 2$, in which case both l and m may have more than one value. The allowed values for the three quantum numbers, when $n = 2$, are :

n	*l*	*m*	*Quantum state*
2	0	0	(200)
2	1	0	(210)
2	1	+1	(211)
2	1	−1	(21-1)

Since the energy of a quantum state depends only on the value of n, the second quantum state is *fourfold degenerate.* Thus

$$E_{(200)} = E_{(210)} = E_{(21-1)} = \frac{E_{(100)}}{4} = -0.545 \times 10^{-18}\text{J}.$$

Higher quantum states will exhibit even greater degeneracies. The degenerate states corresponding to the third energy level (Table 9.1) *ninefold degenerate.*

TABLE 9.1
Quantum States of the Third Energy Level

n	*l*	*m*	*Quantum state*
3	0	0	(300)
3	1	0	(310)
3	1	+1	(311)
3	1	−1	(31-1)
3	2	0	(320)
3	2	+1	(321)
3	2	−1	(32-1)
3	2	+2	(322)
3	2	−2	(32-2)

The stationary-state energy levels generated from Eq. (9.68) account fully for the atomic spectrum of hydrogen.

There are, however, some very small corrections that result from relativistic considerations. These will be accommodated later by incorporation of an additional *electron-spin* quantum number. Furthermore, knowing the number of degenerate quantum states corresponding to any particular stationary-state energy helps one to interpret spectral line intensities.

The interpretation of line intensities is not quite so simple as it may appear from the above discussion. A more thorough quantum-mechanical study of the transitions between quantum states indicates that certain transitions are *forbidden.* Such restrictions are summarized in what are called the *selection rules.* In the case of atomic spectra, the selection rules state that n may change by any integer, l must change by ± 1, and m may change by ± 1 or not at all.

TABLE 9.2

A Summary of the Quantum Numbers

Quantum number	*Name*	*Allowed values*	*Eigenvalue quantized*
n	Principal	$1, 2, 3, \ldots, \infty$	$E = \frac{-2\pi^2 \mu Z^2 e^2}{n^2 h^2}$
l	Azimuthal	$0, 1, 2, \ldots, (n-1)$	$B = l(l+1)$
m	Magnetic	$0, \pm 1, \pm 2, \ldots, \pm l$	$A = -m^2$

THE TOTAL WAVE FUNCTION ψ

The total normalized wave function for relative motion in the hydrogen atom or hydrogenlike ion is expressed, as

$$\psi_{nlm}(r, \theta, \varphi) = R_{nl}(r) \cdot \Theta_{lm}(\theta) \cdot \Phi_m(\varphi), \tag{9.70}$$

where normalized functions for $R_{nl}(r)$ are given by Eq. (9.66), those for $\Theta_{lm}(\theta)$ are given by Eq. (9.31) and those for $\Phi_m(\varphi)$ are given by Eq. (9.17).

It can be easily shown that normalization of the R, Θ, and Φ functions ensures the normalization of the total ψ function. It is interesting to note that just as for the single particle in the three-dimensional box, where ψ was a function of n_x, n_y, and n_z, we have found that *three* quantum numbers n, l, and m are needed in order to define the wave functions for the hydrogen atom. In fact, we may state generally that in a given system, *each degree of freedom requires the introduction of one quantum number.* Where the particle is able to move in any one of three coordinate directions, three quantum numbers arise.

The three quantum numbers which naturally arises in the non-relativistic quantum-mechanical solution of the wave function for relative motion in the hydrogen atom or hydrogen-like ion are summarized in Table 9.2.

ANGULAR MOMENTUM

We have seen that the quantized energy of relative motion of the hydrogen atom is entirely dependent on the value of the principal quantum number *n*. Through use of the operator postulate, it will be clear that the azimuthal quantum number *l* is associated in a similar fashion with the total angular momentum of the atom and that the magnetic quantum number *m* is associated with the orientation off the angular momentum vector in an external magnetic field.

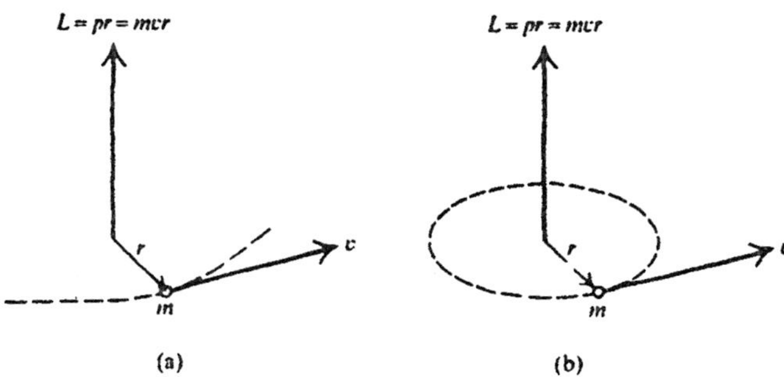

Fig. 9.2. Vector representation of the angular momentum *L* for a mass *m* moving about an axis with a linear velocity *v*, at a radius of curvature *r*. (a) Any curved motion. (b) Circular motion.

Angular Momentum Operators

Whenever a mass *m* moves in a curved path with a velocity *v*, it has an angular momentum *L*, which is given classically as

$$L = mvr = pr, \tag{9.71}$$

where *p* is the linear momentum and *r* is the radius of curvature. In vector algebra the angular momentum is represented by a vector arrow of length *pr*, which is perpendicular to both the radius vector and the linear velocity vector (Fig. 9.2). If the motion is circular, *r* is a constant and the angular momentum vector is represented by an arrow projecting perpendicularly from the center of the circle. It may be shown that the total angular momentum *L* of a system may also be expressed in terms of its vector components L_x, L_y, and L_z in the cartesian coordinate directions *x, y,* and *z,* such that

$$L_x = yp_z - zp_y, \tag{9.72}$$

$$L_y = zp_x - xp_z, \tag{9.73}$$

$$L_z = xp_y - yp_{xy}, \tag{9.74}$$

where p_x, p_y, and p_z are the corresponding linear momenta of the system in the x, y, and z component directions. According to the operator rules we may formulate the quantum-mechanical operators for L_x, L_y, and L_z by simply replacing the p terms in Eqs. (9.72), (9.73), and (9.74) with corresponding linear-momentum operator, so that

$$\hat{L}_x = \frac{h}{i2\pi}\left(y\frac{\partial}{\partial z} - z\frac{\partial}{\partial y}\right) \tag{9.75}$$

$$\hat{L}_y = \frac{h}{i2\pi}\left(z\frac{\partial}{\partial x} - x\frac{\partial}{\partial z}\right) \tag{9.76}$$

$$\hat{L}_z = \frac{h}{i2\pi}\left(x\frac{\partial}{\partial y} - y\frac{\partial}{\partial x}\right) \tag{9.77}$$

By applying the transformation equations we can convert the angular momentum operators given by Eqs. (9.75), (9.76), and (9.77) into forms involving r, θ, and ϕ, so that

$$\hat{L}_x = \frac{h}{i2\pi}\left(-\sin\varphi\frac{\partial}{\partial\theta} - \cot\theta\cos\varphi\frac{\partial}{\partial\varphi}\right) \tag{9.78}$$

$$\hat{L}_y = \frac{h}{i2\pi}\left(\cos\varphi\frac{\partial}{\partial\theta} - \cot\theta\sin\varphi\frac{\partial}{\partial\varphi}\right) \tag{9.79}$$

$$\hat{L}_z = \frac{h}{i2\pi}\frac{\partial}{\partial\varphi}$$

and since, by vector algebra,

$$\hat{L}^2 = \hat{L}_x^2 + \hat{L}_y^2 + \hat{L}_z^2,$$

it can be readily shown that

$$\hat{L}^2 = \frac{-h^2}{4\pi^2}\left[\frac{1}{\sin\theta}\frac{\partial}{\partial\theta}\left(\sin\theta\frac{\partial}{\partial\theta}\right) + \frac{1}{\sin^2\theta}\frac{\partial^2}{\partial\varphi^2}\right]. \tag{9.81}$$

It is convenient to express the operator for L_x^2. From Eq. (9.80) we can write

$$\hat{L}_x^2 = \frac{-h^2}{4\pi^2}\frac{\partial^2}{\partial\varphi^2}. \tag{9.82}$$

Orbital Angular Momentum and the Azimuthal Quantum Number *l*

If we multiply Eq. (9.9) by $(h^2\Phi/4\pi^2)$ we obtain

$$\frac{-h^2}{4\pi^2}\left[\frac{1}{\sin\theta}\cdot\frac{\partial}{\partial\theta}\left(\sin\theta\frac{\partial\Theta\Phi}{\partial\theta}\right)+\frac{A\Phi\Theta}{\sin^2\theta}\right]=\left(\frac{bh^2}{4\pi^2}\right)\Theta\Phi. \qquad (9.83)$$

Substitution for $A\Phi$ from Eq. 99.6) and of $l(l+1)$ for B gives

$$\frac{-h^2}{4\pi^2}\left[\frac{1}{\sin\theta}\frac{\partial}{\partial\theta}\left(\sin\theta\frac{\partial\Theta\Phi}{\partial\theta}\right)+\frac{\Theta d^2\Phi}{\sin^2\theta\varphi^2}\right]=\left[\frac{l(l+1)h^2}{4\pi^2}\right]\Theta\Phi. \qquad (9.84)$$

Since Θ is not a function of ϕ, the second term in brackets on the left may be rewritten to include Θ as part of the function with respect to which the ϕ derivative is taken. Thus we have

$$\frac{-h^2}{4\pi^2}\left[\frac{1}{\sin\theta}\frac{\partial}{\partial\theta}\left(\sin\theta\frac{\partial\Theta\Phi}{\partial\theta}\right)+\frac{1}{\sin^2\theta}\frac{\partial^2\Theta\Phi}{\partial\varphi^2}\right]=\left[\frac{l(l+1)h^2}{4\pi^2}\right]\Theta\Phi, \qquad (9.85)$$

or, in operator symbolism,

$$\frac{-h^2}{4\pi^2}\left[\frac{1}{\sin\theta}\cdot\frac{\partial}{\partial\theta}\left(\sin\theta\frac{\partial}{\partial\theta}\right)+\frac{1}{\sin^2\theta}\cdot\frac{\partial^2}{\partial\varphi^2}\right]\Theta\Phi=\left[\frac{l(l+1)h^2}{4\pi^2}\right]\Theta\Phi, \qquad (9.86)$$

But, according to Eq. (9.81), the operator in Eq. (9.86), which is an eigenvalue equation, is $\hat{L}^2$. Substitution of Eq. (9.81) into Eq. (9.86) gives

$$\hat{L}^2\Theta\Phi=\left[\frac{l(l+1)h^2}{4\pi^2}\right]\Theta\Phi, \qquad (9.87)$$

Now since the operator $\hat{L}^2$ does not differentially operate on r, and since $R(r)$ does not depend on θ or on ϕ, we can introduce R into the operand on both sides of Eq. (9.87) in such a way that

$$\hat{L}^2\Theta\Phi R=\left[\frac{l(l+1)h^2}{4\pi^2}\right]\Theta\Phi R, \qquad (9.88)$$

and since the total wave function ψ is equal to $\Theta\Phi R$, we can write

$$\hat{L}^2\psi=\left[\frac{l(l+1)h^2}{4\pi^2}\right]\psi, \qquad (9.89)$$

Equation (9.89) is an eigenvalue equation in which the operator is $\hat{L}^2$, the operand is ψ, the wave function for the hydrogen atom, and the eigenvalues, according to Postulate 3, are the allowed stationary-state values of L^2, the square of the orbital angular momentum of the hydrogen atom. Since l is restricted to integral values, L^2 and L are then quantized such that

$$L^2 = l(l+1)\frac{h^2}{4\pi^2}, \tag{9.90}$$

and

$$L = \sqrt{l(l+1)}\,\frac{h}{2\pi}, \tag{9.91}$$

where $l = 0, 1, 2, ..., (n-1)$. The values of L calculated from Eq. (9.91) are the **only exact values of L which may be observed experimentally.** The negative square root was not included in writing Eq. (9.91) because negative magnitude for vectors are meaningless.

Thus, just as the principal quantum number *n* serves to quantize the allowed energy of relative motion of the hydrogen atom, the azimuthal quantum number *l* serves to quantize the allowed levels for the orbital angular momentum of the hydrogen atom.

Compare this to the Bohr model in which the angular momentum of the atom was quantized by *postulation* as

$$L = \mu vr = n\frac{h}{2\pi}. \tag{9.92}$$

It is important to note that the Sommerfeld extension of the Bohr theory required quantization of angular momentum according to Eq. (9.91), Eq. (9.92). Moreover, in the Schrodinger model, the principal quantum number is related to the *energy* of relative motion and not directly to the angular momentum.

The relationship between angular momentum and magnetic properties can be easily understood by imagining an idealized model in which the electron moves in a classical plannar orbit, as shown in Fig. 9.3. For the direction of electron revolution indicated, the angular monrentum vector L of length $\sqrt{l(l+1)}\,h/2\pi$ is pointed upward. Whenever a charged particle revolves about an axis, a magnetic field is generated along that axis. Such a device is called an electromagnet. In determining the direction of the magnetic field, *when using electrons,* we use a "left-hand rule" such that if the curled fingers of the left hand point in the direction of the electron flow around the axis, the left thumb points toward the *north pole* of the electromagnet. Thus, in Fig. 9.3, the magnetic moment vector μ_l is pointed in a direction exactly opposite to that of the orbital angular momentum vector *L*. It can be shown that the magnitude of the magnetic moment vector is also quantized by the azimuthal quantum number *l*.

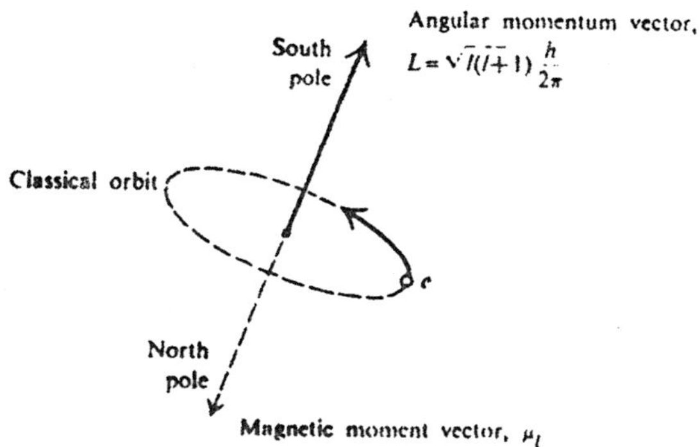

Fig. 9.3. A simplified pictorialization of the relationship between the orbital angular momentum vector and the magnetic moment vector for an electron in a classical orbit.

It is customary to designate *l* quantum states by letters, according to the following scheme :

Value of *l*	0	1	2	3	4	5 ...
Orbital notation	s	p	d	f	g	h ...

Thus, when the electron in a quantum state in which $n = 1$ and $l = 0$, it is said to be in the 1s *orbital;* when it is in a state in which $n = 3$ and $l = 1$, it is said to be in the 3p orbital. **We define an orbital as a one- electron wave function. The letters s, p, d, and f are derived from the spectroscopic terms : sharp, principal, diffuse, and fundamental.**

It is interesting to note that for all sates in which $l = 0$, that is, for all s orbitals, the orbital angular momentum has a value of zero according to Eq. (9.91).

The *z*-component of Angular Momentum and the Magnetic Quantum number *m*

We have defined the allowed stationary-state values for the orbital angular momentum of the electron according to Eq. (9.91). However, we are unable to state the *orientation* of the angular-momentum vector with respect to an external reference. If there is no external magnetic field, the angular momentum vector L, which is directed oppositely to the magnetic moment vector μ_l, should not be constrained to any particular orientation. However, in the presence of an external magnetic field, different orientations of the magnetic moment vector, and the associated antiparallel angular momentum vector, with respect to the

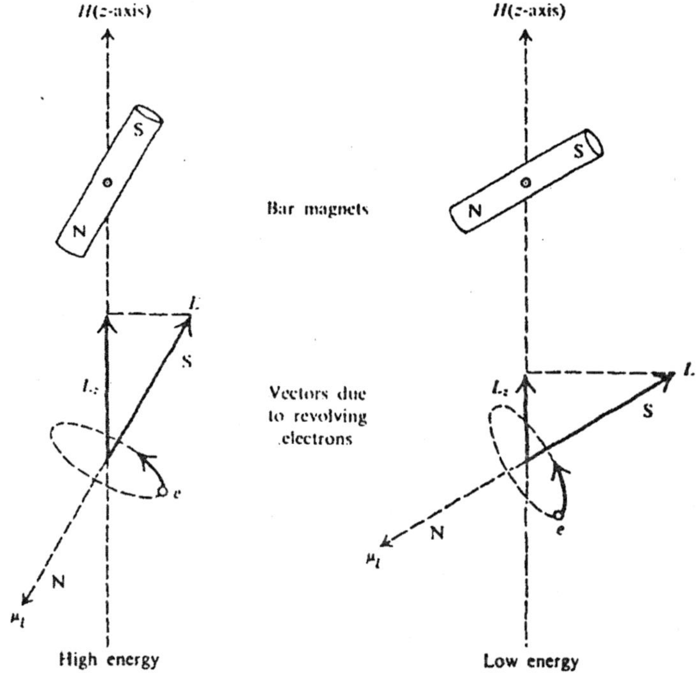

Fig. 9.4. Relations between orientation and energy for angular momentum vectors L and magnetic moment vectors μ_l, in the presence of an external magnetic field H.

external field direction represent different energies. For illustration, consider the two orientations of the angular momentum vector and its associated magnetic moment vector illustrated in Fig. 9.4. To simplify the representation planar orbits for the electrons have been shown. Let us impose an *external magnetic field H* on the system and define the z-axis as the field direction. By convention, the field arrow H points in the direction *toward which an imaginary isolated north pole would migrate.* Thus, the more nearly the arrow of the magnetic moment vector μ_l aligns itself with the field direction, the lower will be the energy of the orientation. In terms of L, the more nearly the angular momentum vector L aligns itself with the external magnetic field direction, the ***higher*** will be the energy of the system. In order to compare, two pivoted bar magnets in the same relative orientations as the two vectors are also shown in fig. 9.4. A convenient measure of the orientation of the vector with respect to the z-axis is the length of the z-component, L_z, of the angular momentum vector. The component L_z is the projection of the L vector on the z-axis. In Fig. 9.4, although L is identical in the high-energy and low-energy orientations, L_z is larger in the higher-energy orientation.

Let us show that the energy related to the ***orientation*** of the angular momentum vector in an external field is quantized. (More specifically L_z is quantized). On multiplying both sides of Eq. (9.6), the eigenvalue equation for $\Phi(\phi)$, by $(-h^2/4\pi^2)$ we get

$$\left(\frac{-h^2}{4\pi^2}\frac{d^2}{d\varphi^2}\right)\Phi = \left(\frac{-4^2A}{4\pi^2}\right)\Phi. \tag{9.93}$$

The operator in Eq. (9.93) is $\hat{L}_z^2$, as given by Eq. (9.82). Substitution of Eq. (9.82) and $(-m^2)$ for A into Eq. (9.93) yields

$$L_z^2\Phi = \frac{m^2h^2}{4\pi^2}\Phi. \tag{9.94}$$

Since the operator $\hat{L}_z^2$ does not differentially operate on θ or r, and since neither $\Theta(\theta)$ nor $R(r)$ contains ϕ, it is possible to introduce ΘR into the operand on both sides of Eq. (9.94) in such a way that

$$\hat{L}_z^2\Phi\Theta R = (m^2h^2/4\pi^2)\Phi\Theta R,$$

and since $\psi = \Phi\Theta R$, we can write

$$\hat{L}_z^2\psi = (m^2h^2/4\pi^2)\psi. \tag{9.95}$$

Equation (9.95) is an eigenvalue equation in which the operator is $\hat{L}_z^2$, the operand is ψ, the wave function for the hydrogen atom, and the eigenvalues, according to the operator postulate, are the allowed stationary- state values for L_z^2, the square of the z- component of angular momentum. Since m is restricted to integral values, L_z^2 and L_z are each quantized such that

$$L_z^2 = m^2 \frac{h^2}{4\pi^2}, \tag{9.96}$$

and

$$L_z = m \frac{h}{2\pi}, \tag{9.97}$$

where $m = -l\, -(l - 1), \ldots, 0, \ldots, (l - 1), l.$

The *only exact experimental values* for L_z which may be observed are those calculated from Eq. (9.97). The negative square root in Eq. (9.97) is also a satisfactory solution, but this is accommodated by the allowed negative values for the integer m. thus, **the magnetic quantum number m serves to quantize the allowed levels for the z-component of angular momentum.** That is, the quantum number m serves to quantize the orientation of the orbital angular momentum vector in the presence of an external magnetic field.

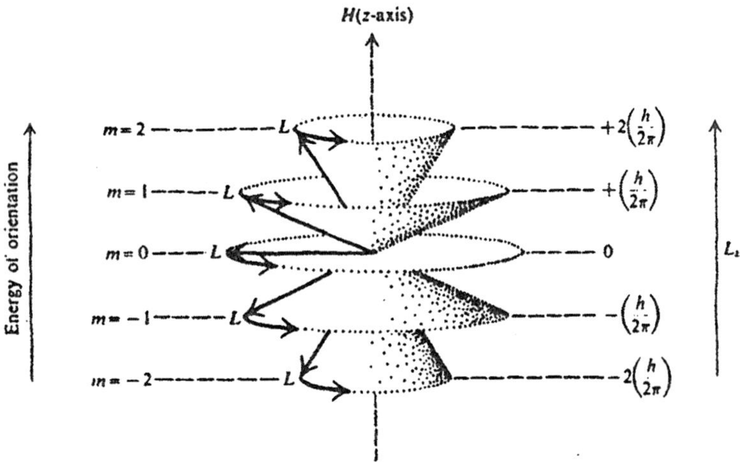

Fig. 9.5. Permitted orientations of the angular momentum vector L for a electron (l = 2) in an external magnetic field. The L vectors, each of length $\sqrt{2(1 + 1)}\, h/2\pi$, precess about the z-axis.

The statement of n, l, and m for the quantum state of an electron in the hydrogen atom allows the immediate calculation of its total energy,

$$E = -2\pi^2 me^4/n^2h^2,$$

its orbital angular momentum, $L = \sqrt{l(l+1)}\,h/2\pi$,

and the z-component of its angular momentum, $L_z = mh/2\pi$.

According to quantum mechanics, the values calculated from the above equations are the **only possible observable exact values** for E, L, and L_z.

The allowed orientations of the angular momentum vector for a d electron ($l = 2$) are shown in Fig. 9.5. The length of the vector in all class is $L = \sqrt{(2(2+1)}\,h/2\pi = \sqrt{6}\,h/2\pi$. Since $l = 2$, m may assume the values –2, –1, 0, 1, and 2, and L_z may thus have the five values, according to Eq. (9.97)

$$-2\left(\frac{h}{2\pi}\right), -\left(\frac{h}{2\pi}\right), 0, +\left(\frac{h}{2\pi}\right), +2\left(\frac{h}{2\pi}\right).$$

The Hydrogen-like Wave Functions

Some of the wave functions for the hydrogen atom are given in Table 9.1.

TABLE 9.1
NORMALIZED HYDROGEN-LIKE WAVE FUNCTIONS FOR THE FIRST TWO ENERGY LEVELS

$$\psi_{nlm}(r, \theta, \varphi) = R_{nl}(r)\cdot\Theta_{lm}(\theta)\cdot\Phi_m(\varphi)$$

n	l	m	$R_{ln}(r)$ (radial function)	$\Theta_{lm}(\theta)\cdot\Phi_m(\varphi)$ (sperical harmonic)	Symbol for wave function or orbital
1	0	0	$2\left(\frac{z}{a_0}\right)^{3/2} e^{-Zr/2a_0}$	$\left(\frac{1}{4\pi}\right)^{1/2}$	1s
2	0	0	$\left(\frac{z}{2a_0}\right)^{3/2}\left(2-\frac{Zr}{a_0}\right)e^{-Zr/2a_0}$	$\left(\frac{1}{4\pi}\right)^{1/2}$	2s
2	1	0	$\frac{1}{\sqrt{3}}\left(\frac{z}{2a_0}\right)^{3/2}\left(\frac{z\pi}{2a_0}\right)e^{-Zr/2a_0}$	$\left(\frac{3}{4\pi}\right)^{1/2}\cos\theta$	$2p_0$ or $2p_z$
2	1	+1	$\frac{1}{\sqrt{3}}\left(\frac{z}{2a_0}\right)^{3/2}\left(\frac{Zr}{2a_0}\right)e^{-Zr/2a_0}$	$\left(\frac{3}{8\pi}\right)^{1/2}\sin\theta e^{i\varphi}$	$2p_{+1}$
2	1	-1	$\frac{1}{\sqrt{3}}\left(\frac{z}{2a_0}\right)^{3/2}\left(\frac{Zr}{2a_0}\right)e^{-Zr/2a_0}$	$\left(\frac{3}{8\pi}\right)^{1/2}\sin\theta e^{-i\varphi}$	$2p_{-1}$

EQUILALENT HYBRID FUNCTIONS FOR $2p_{+1}$ and $2p_{-1}$

2	1	±1 hybrid	$\frac{1}{\sqrt{3}}\left(\frac{z}{2a_0}\right)^{3/2}\left(\frac{Zr}{2a_0}\right)e^{-Zr/2a_0}$	$\left(\frac{3}{4\pi}\right)^{1/2}\sin\theta\cos\varphi$	$2p_x$
2	1	±1 hybrid	$\frac{1}{\sqrt{3}}\left(\frac{z}{2a_0}\right)^{3/2}\left(\frac{Zr}{2a_0}\right)e^{-Zr/2a_0}$	$\left(\frac{3}{4\pi}\right)^{1/2}\sin\theta\sin\varphi$	$2p_y$

Normalized Spherical Harmonics for d Orbitals (l = 2)

l	m	$\Theta_{lm}(\theta)\Phi_m(\varphi)$	*Symbol*
2	0	$\left(\frac{5}{16\pi}\right)^{1/2}(3\cos^2\theta - 1)$	d_0 or d_{z^2}
2	+1	$\left(\frac{15}{8\pi}\right)^{1/2}\sin\theta\cos\theta\, e^{i\varphi}$	d_{+1}
2	−1	$\left(\frac{15}{8\pi}\right)^{1/2}\sin\theta\cos\theta\, e^{-i\varphi}$	d_{-1}
Equivalent Hybrid Functions for d_{+1} and d_{-1}			
2	±2 hybrid	$\left(\frac{15}{4\pi}\right)^{1/2}\sin\theta\cos\theta\cos\varphi$	d_{xz}
2	±2 hybrid	$\left(\frac{15}{4\pi}\right)^{1/2}\sin^2\theta\cos\theta\sin\varphi$	d_{yz}
2	+2	$\left(\frac{15}{32\pi}\right)^{1/2}\sin\theta\, e^{i2\varphi}$	d_{+2}
2	−2	$\left(\frac{15}{32\pi}\right)^{1/2}\sin\theta\, e^{-i2\varphi}$	d_{-2}
Equivalent Hybrid Functions for d_{+2} and d_{-2}			
2	±2 hybrid	$\left(\frac{15}{16\pi}\right)^{1/2}\sin^2\theta\sin 2\varphi$	d_{xy}
2	±2 hybrid	$\left(\frac{15}{16\pi}\right)^{1/2}\sin^2\theta\cos 2\varphi$	$d_{x^2-y^2}$

For the sake of convenience, the total wave function

$$\psi_{nlm}(r, \theta, \varphi) = R_{nl}(r)\cdot\Theta_{lm}(\theta)\cdot\Phi_m(\varphi) \tag{9.98}$$

is presented as the product of two parts : *a radial function,* $R_{nl}(r)$, and an *angular function,* $\Theta_{lm}(\theta)\Phi_m(\varphi)$.

The radial function $R_{nl}(r)$ depends only on r has the same form for all states which have the same values for their n quantum numbers and for their l quantum numbers. All of the 2p orbitals, for instance, have the same form for $R(r)$.

The angular part of the wave function, $\Theta_{lm}(\theta) \cdot \Phi_m(\varphi)$, is also called the **spherical harmonic** and shows the combined dependence of ψ on θ and ϕ.

SHAPES OF ORBITALS

Orbitals

It is clear from Table 9.1 that the spherical harmonics $\Theta(\theta)\Phi(\phi)$ for the 1s and 2s orbitals are not dependent on the values of θ and

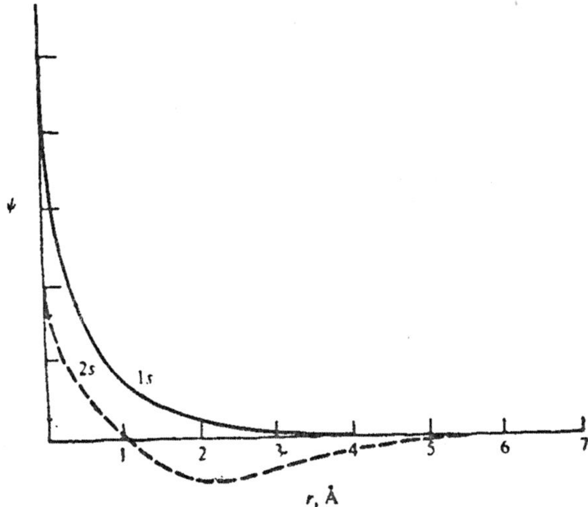

Fig. 9.6. The ψ wave functions for the electron in the 1s orbital and in the 2s orbital of the hydrogen atom, ψ is given in relative units.

ϕ. In addition, it follows from Eq. (9.98) that the *total* wave function ψ is independent on both θ and ϕ for the 1s and 2s states. Furthermore, independence of θ and ϕ may be shown for the wave functions of *all* s orbitals, that is, the 3s orbital, the 4s orbital, and so on. Thus, the wave functions ψ_{1s}, ψ_{2s}, ψ_{3s}, ... or, 1s, 2s, 3s, ..., depend only on r and are spherically symmeterical. Plots of ψ versus r for the 1s orbital and the 2s orbital are shown in Fig. 9.6. Since, for s orbitals, $R(r)$ is directly proportional to ψ, a corresponding plot of $R(r)$ versus r would have the same form as that for the ψ versus r curves shown in Fig. 9.6.

The plots of ψ versus r or $R(r)$ versus r do not provide much useful information. We are generally more interested in the probability distribution of the electron about the nucleus, which is given by plots of $\psi^*\psi$ versus r, or ψ^2 versus r. Probability density distributions, ψ^2 versus r, are given for the 1s and 2s orbitals in Fig. 9.7. ψ^2 is the probability *per unit volume* of finding the electron at a given point in space. For both the 1s orbital and the 2s orbital, ψ^2 is maximum in the center of the atom. Three-dimensional plots of ψ^2 versus r are shown as cloud density patterns and boundary surface plots in Fig. 9.8 for 1s and 2s orbitals

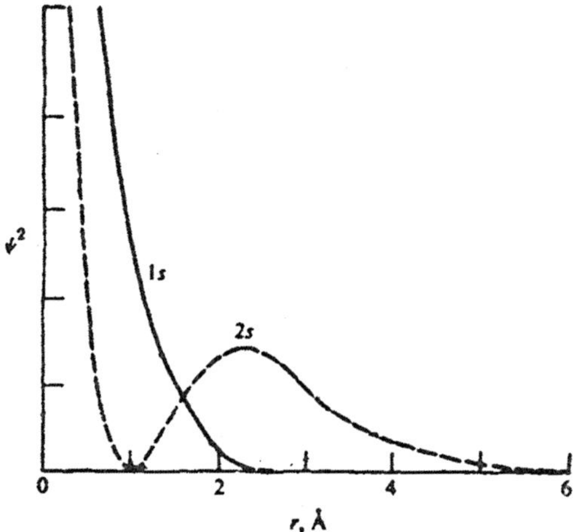

Fig. 9.7. Probability density (ψ^2) distribution for the electron in the 1s and 2s orbitals of the hydrogen atom.

In locating electrons in atoms, we are concerned with the probability of finding the electron at some particular value of *r irrespective of direction*, rather than at a particular value of r, θ, and ϕ. Specifically, we calculate the probability that an electron exists between a sphere of radius r and a concentric sphere of radius $r + dr$, that is, the probability that it exists within the annular shell of volume $4\pi r^2 dr$. Such a *radial probability* is given by the product of the

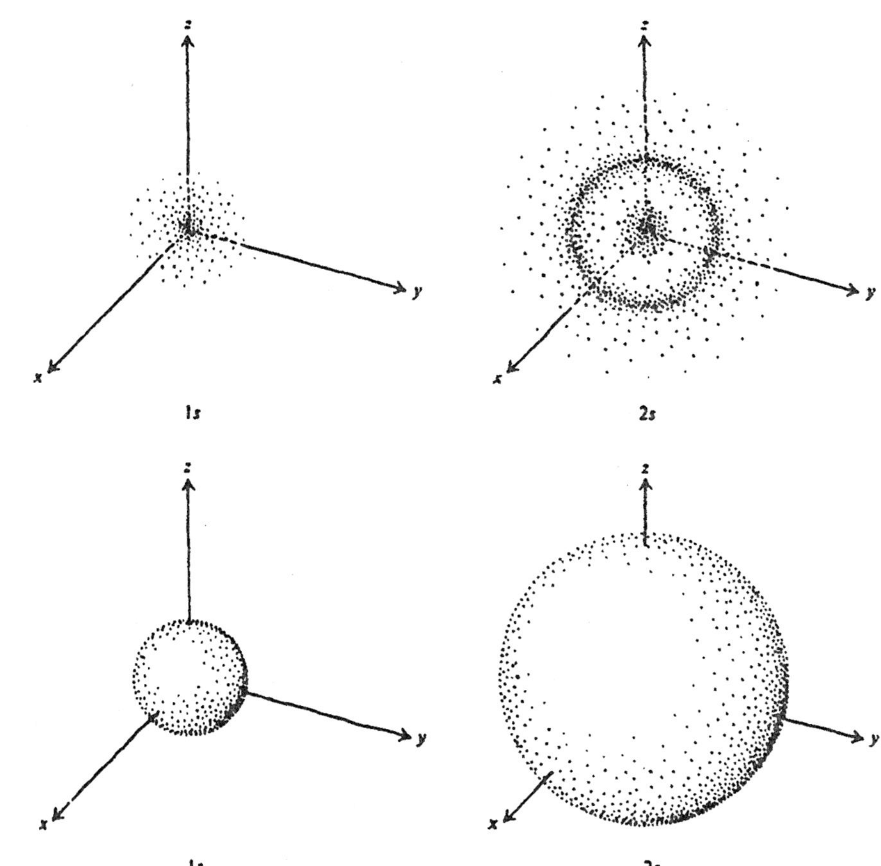

Fig. 9.8. Probability density (ψ^2) as a function of position for the electron in the 1s and 2s orbitals of hydrogen. (a) A cross section in which cloud density is proportional to ψ^2. (b) A boundary surface which excludes all space elements in which ψ^2 is less than one-tenth of the maximum value.

volume of the shell, $4\pi r^2\, dr$ (surface area × thickness) multiplied by ψ^2, the probability per *unit* volume that the electron exists at the same value of r. The radial probability is then given as

$$4\pi r^2 \psi^2\, dr.$$

For all s orbitals

$$\psi = R(r)(1/4pi)^{1/2},$$

Radial probability thus becomes

$$r^2R^2(r)\,dr.$$

This gives the probability of finding the electron in a shell of thickness dr at a distance r from the nucleus. A *finite* probability may be expressed if we define the *radial distribution function* $D(r)$ as the probability *per unit radius* of finding the electron in a spherical shell at a distance r from the nucleus. Thus

$$D(r) = r^2R^2(r)\,dr/dr = r^2R^2(r).$$

Plots of the radial distribution function $r^2R^2(r)$ for the 1s and 2s orbitals are showing in Fig. 9.9. For the 1s orbital, which represents the ground state, the most probable radius is given as a_0 = 52.9 pm, **which is exactly the same as that calculated from the Bohr theory.**

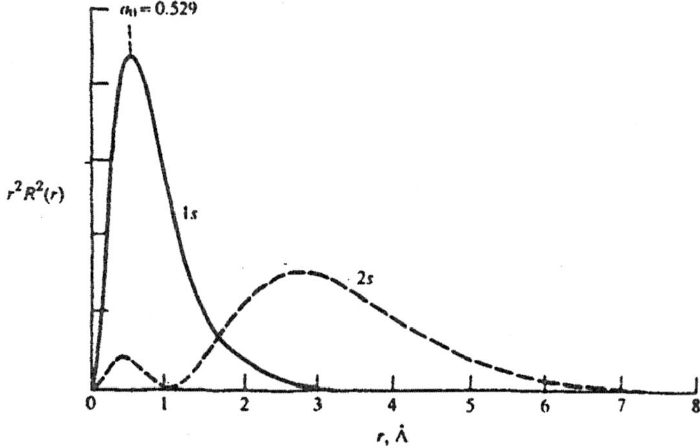

Fig. 9.9. Radial distribution functions $r^2R^2(r)$ for the electon in the 1s and 2s orbitals of hydrogen.

Thus, the quantum-mechanical model for the hydrogen atom allows the electron to have some probability for existence, however, small, at *all* values of r. However, there is a maximum probability per unit radius for finding the electron at $r = a_0$.

For the 2s orbital there are two maxima or peaks in the radial distribution curve : one just below 300 pm and a much smaller peak at about 40 pm. The unexpected appearance of the smaller peak much closer to the nucleus is referred to as *penetration*. In multi-electronic

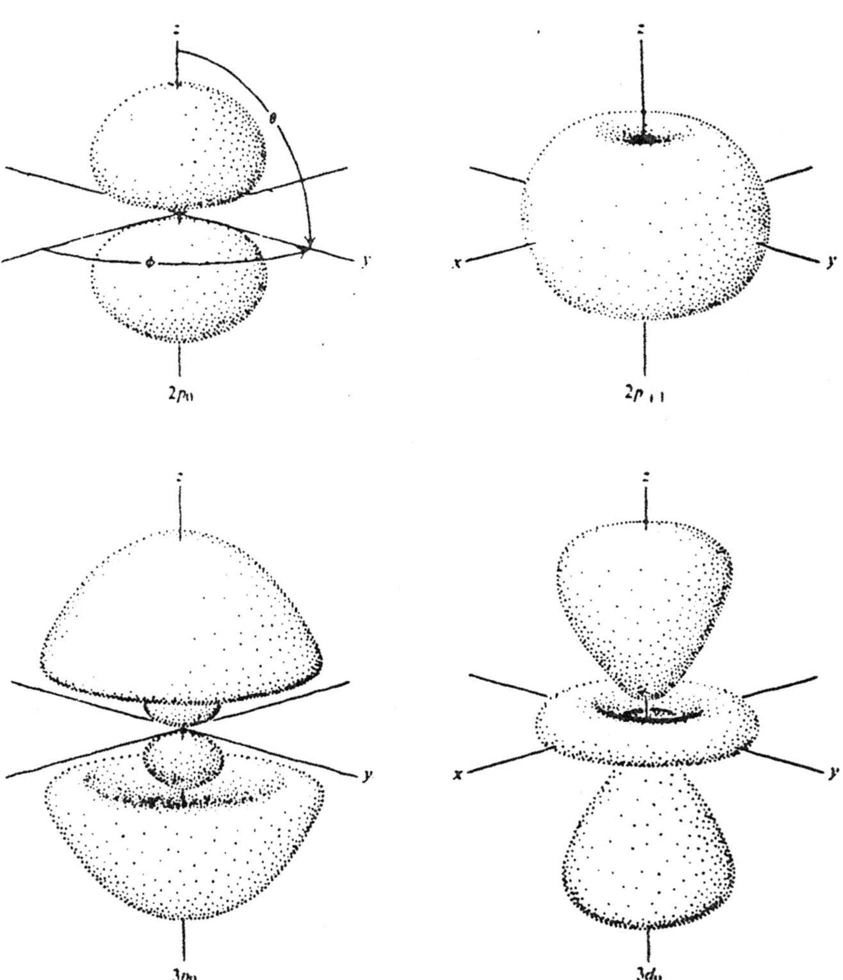

Fig. 9.10. Boundary surface plots of the probability distributions ($\psi^*\psi$ versus r, θ, and ϕ) for several orbitals with angular momentum. The boundary surfaces exclude all space in which the probability density is less than one-tenth of the maximum.

atoms, penetration effects are important in states with low values of l and contribute significantly to the bonding energy between the electron and the nucleus. States with high penetration have low energy and high stability.

p-Orbitals

Table 9.1 indicates that the ψ functions for those states for which l is greater than zero are not symmeterical about the nucleus but depend on the values of θ and ϕ. States for which $l > 0$ are states which have angular momentum. For example, for the (210) quantum state, or the $2p^0$ orbital, ψ depends on r and on θ, but not on ϕ, and is thus symmetrical about the z-axis.

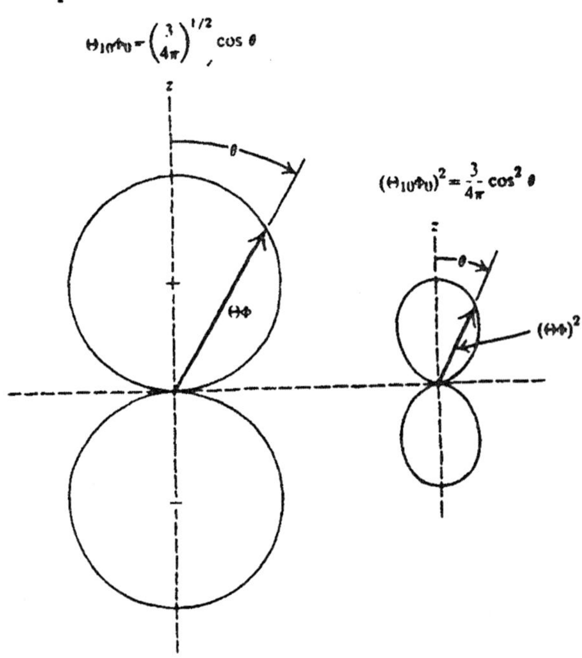

Fig. 9.11. Two-dimensional polar graphs of $\Theta\Phi$ versus θ and $(\Theta\Phi)^2$ versus θ for P_0 or P_z orbitals. Since $\Theta\Phi$ is independent of ϕ for p_0 orbitals, the two- dimensional plots appear the same for any value of ϕ. The values of $\Theta\Phi$ and $(\Theta\Phi)^2$ are given by the length of the radial coordinate.

Boundary-surface plots of $\psi^*\psi$ versus r, θ, and ϕ for several orbitals having angular momentum are given Fig. 9.10. The contours shown are considerably more complicated than those for the 1s and 2s orbitals. The probability term plotted in Fig. 9.10 is

$$\psi^*\psi = |\psi|^2 = |R|^2\,|\Theta\Phi|^2,$$

or since none of the radial functions are complex,

$$\psi^*\psi = R^2\,|\Theta\Phi|^2. \tag{9.100}$$

Since $\Phi_m(\theta)$ is a complex function for all values of m other than zero, we must express all spherical harmonics for which $m \neq 0$ in

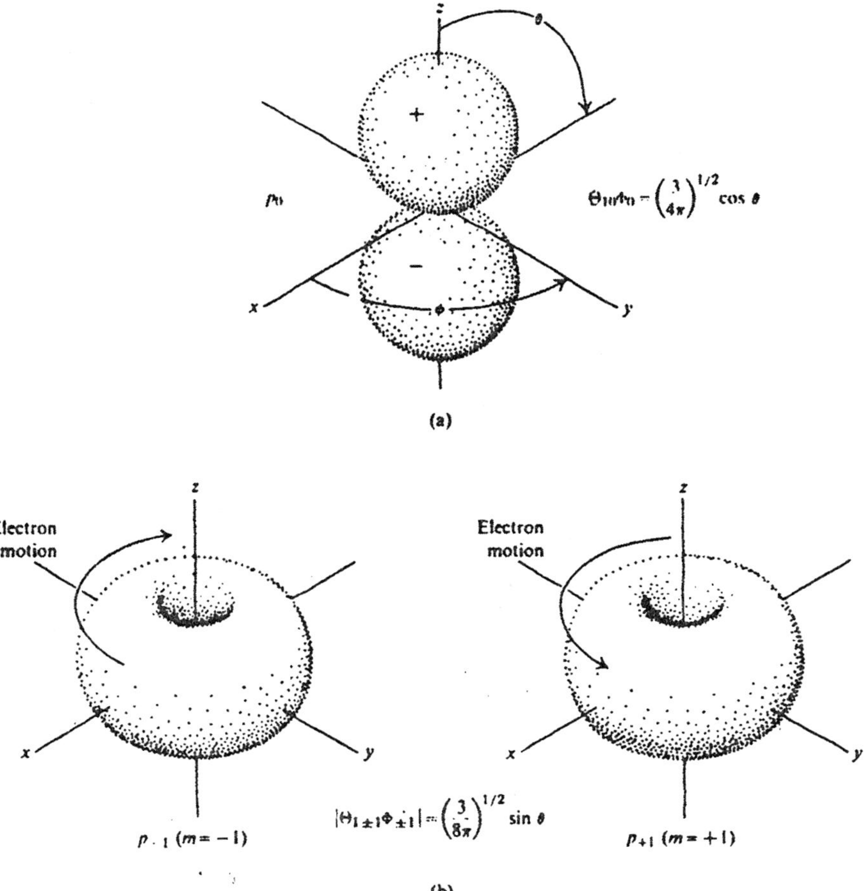

Fig. 9.12. (a) Spherical-polar graph of $\Theta\Phi$ versus θ and ϕ for the p_0 orbitals. The value of $\Theta\Phi$ is given by the length of the radial coordinate. (b) Spherical-polar graphs of the absolute values $|\Theta\Phi|$ of the spherical harmonics for p_{+1} and p_{-1} orbitals.

terms of their absolute values $|\Theta\Phi|$ if we are to use them in Eq. (9.100). Since the R^2 term in Eq. is spherically symmetrical, it is more convenient to consider *angular* dependence of the probability distribution in terms of the square of the spherical harmonic $(\Theta\Phi)^2$ or the square of the *absolute* spherical harmonic $|\Theta\Phi|^2$ in the case of complex functions. In addition, the spherical harmonic $\Theta\Phi$ or the absolute spherical harmonic $|\Theta\Phi|$ is often used without squaring for plots of angular dependence of wave functions. The advantage of using plots involving only the spherical harmonic and not R is that the forms for $\Theta\Phi$ are not dependent on the value of n, the principal quantum number. Thus, graphs of $|\Theta\Phi|$ versus θ and ϕ and of $|\Theta\Phi|^2$ versus θ and ϕ are the same for all p_0 orbitals (regardless of n), for all P_{+1} orbitals, for all p_{-1}, for all d_0 orbitals, and so on.

Two-dimensional polar graphs of $\Theta\Phi$ versus θ and of $(\Theta\Phi)^2$ versus θ are shown in Fig. 9.12 for the p_0 orbitals, which are not complex. Since $\Theta\Phi$ and $(\Theta\Phi)^2$ for p_0 orbitals are independent of ϕ, the corresponding three-dimensional graphs of $\Theta\Phi$ versus θ and ϕ and of $(\Theta\Phi)^2$ versus θ and ϕ are symmetric about the z-axis, as shown for the spherical polar plot of $\Theta\Phi$ versus θ and ϕ given in Fig. 9.12(a). Note that in a *polar* graph the value of $\Theta\Phi$ is given as the length of the radial coordinate. Note especially that the surfaces shown in Fig. 9.12 are *not* boundary surfaces such as those presented in Fig. 9.12. Spherical polar graphs of $|\Theta\Phi|$ versus θ and ϕ are shown in Fig. 9.12(b) for the spherical harmonics corresponding to the p_{+1} and p_{-1} orbitals. It may be easily shown that the *absolute* values of the spherical harmonics for the p_{+1} and p_{-1} orbitals are identical and are given as

$$|\Theta_{\pm 1}\Phi_{+-1}|' = (3/8\pi)^{1/2} \sin\theta.$$

Thus, for the p_{+1} and p_{-1} orbitals, the three-dimensional spherical polar graphs of $|\Theta\Phi|$ versus θ and ϕ are identical (Fig. 9.12(b)). Each of the plots is symmetric about the z-axis, since $|\Theta\Phi|$ is independent of ϕ for p_{+1} and p_{-1} orbitals. The only distinction we can make between the p_{+1} orbitals and the p_{-1} orbital is with respect to the direction of classical rotation of the electron about the z-axis.

d-Orbitals

Normalized spherical harmonics for d orbitals, are presented in Table 9.1. For a d orbitals, $l = 2$, so that m is allowed the values -2, -1, 0, $+1$, $+2$. A spherical polar graph of the spherical harmonic $\Theta\Phi$ for the d_0 orbital is given in Fig. 9.13 (a), and spherical polar graphs of the *absolute* spherical harmonics $|\Theta\Phi|$ are given for the d_{+1}, d_{-1}, d_{+2} and d_{-2} orbitals in Fig. 9.13 (b) and (c).

As in the case of the p orbitals, the d_{+1} and d_{-1} plots are indistinguishable and are shown as a single $d_{\pm 1}$ plot. In addition, the d_{+2} and d_{-2} graphs are also indistinguishable and are shown as a single $d_{\pm 2}$ plot.

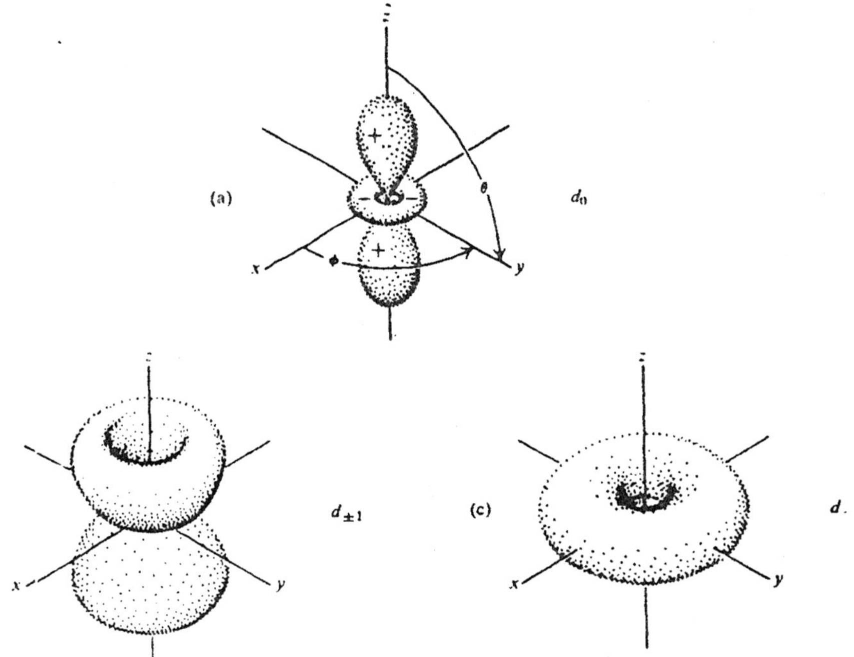

Fig. 9.13. Spherical harmonics for d orbitals. (a) $\Theta\Phi$ versus θ and ϕ for the d_0 orbital. (b) $|\Theta\Phi|$ versus θ and ϕ for the d_{+1} and d_{-1} orbitals. (c) $|\Theta\Phi|$ versus θ and ϕ for the d_{+2} and d_{-2} orbitals.

HYBRID ORBITALS

In order to interpret the geometry of molecules we can construct a new but equivalent set of *real* orthonormal wave functions through linear combination. Since the $2p_{+1}$ and $2p_{-1}$ orbitals are degenerate, any linear combination of these two wave functions will also be an eigenfunction which satisfies the Schrodinger equation.

It is convenient to define the following two new linear combinations of $2p_{+1}$ and $2p_{-1}$:

$$2p_x = \frac{1}{\sqrt{2}}(2p_{+1} + 2p_{-1}) \tag{9.101}$$

and

$$2p_y = \frac{-i}{\sqrt{2}}(2p_{+1} + 2p_{-1}) \tag{9.102}$$

The $2p_x$ function and the $2p_y$ function are called *hybrid* functions. The particular linear combinations given by Eqs. (9.101) and (9.102) are chosen to provide *real orthonormal* hybrid functions with convenient directional properties. The original $2p_{+1}$ and $2p_{-1}$ wave functions (Table 9.1) are

$$2p_{+1} = R_{21}(r)(3/8\pi)^{1/2} \sin \theta e^{i\varphi}, \tag{9.103}$$

and

$$2p_{-1} = R_{21}(r)(3/8\pi)^{1/2} \sin \theta e^{-i\varphi}, \tag{9.104}$$

Substitution of Eqs. (9.103) and (9.104) into Eq. (9.101) yields

$$2p_x = \tfrac{1}{2} R_{21}(r)(3/'r\pi)^{1/2} \sin \theta (e^{i\varphi} + e^{-i\varphi}). \tag{9.105}$$

But, according to Euler's formula,

$$e^{i\varphi} = \cos \varphi + i \sin \varphi, \tag{9.106}$$

and

$$e^{-i\varphi} = \cos \varphi - i \sin \varphi, \tag{9.107}$$

so that

$$e^{i\varphi} + e^{-i\varphi} = 2 \cos \varphi. \tag{9.108}$$

Substitution of Eq. (9.108) into Eq. (9.105) give,

$$2p_x = R_{21}(r)(3/4\pi^{1/2} \sin \theta \cos \varphi.$$

Thus, for any p_x orbitals, for *any* value of *n*, the spherical harmonic is

$$\Theta\Phi)_{p_x} = (3/4\pi^{1/2} \sin \theta \cos \varphi. \tag{9.109}$$

The second hybrid function $2p_y$ is found by substitution of Eqs. (9.103) and (9.104) into Eq. (9.102) :

$$2p_y = -i\tfrac{1}{2}R_{21}(r)(3/r\pi)^{1/2}\sin\theta(e^{i\varphi} + e^{-i\varphi}). \qquad (9.110)$$

Combination of Eqs. (9.106) and (9.107) gives

$$e^{i\varphi} + e^{-i\varphi} = 2i\sin\varphi. \qquad (9.111)$$

and substitution of Eq. (9.111) into Eq. (9.110) yields

$$2p_y = R_{21}(r)(3/4\pi^{1/2}\sin\theta\sin\varphi. \qquad (9.112)$$

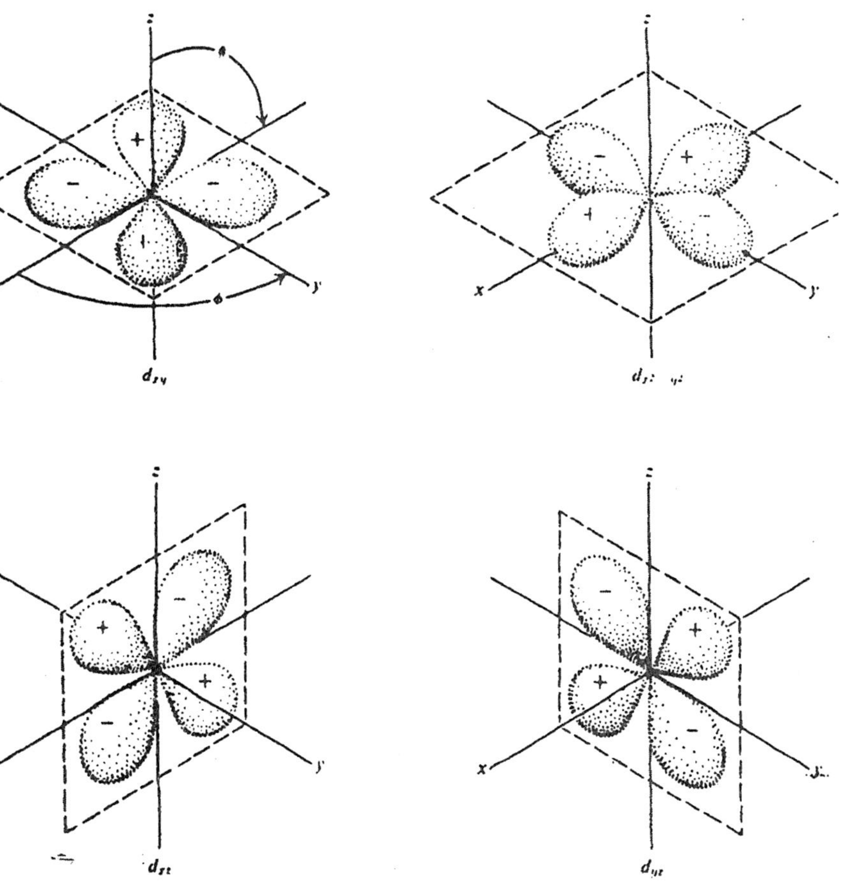

Fig. 9.14. The spherical harmonics ΘΦ for the hybrid d orbitals. In each case, the lobes are bisected by the dotted planes.

The spherical harmonic for the p_y hybrid orbital, for *any* value of *n* may therefore be written as

$$(\Theta\Phi)_{p_y} = (3/4)\pi^{1/2} \sin\theta \sin\varphi. \tag{9.113}$$

For geometric interpretation, we may develop new linear combinations (as for the $p_{\pm 1}$ orbitals) in order to define *two* new *real* and *orthonormal* hybrid functions *each* for the $d_{\pm 1}$ and $d_{\pm 2}$ orbitals.

The real spherical harmonics for the four resulting hybrid orbitals, d_{xz}, d_{yz}, d_{xy}, and $d_{x^2-y^2}$, are given in Table 9.1 and spherical polar plots are shown in Fig. 9.14. The d_0 orbital is not changed; that is, it is not hybridized. It is, however, given the new designation d_{z^2}. It is easy to visualize the spatial equivalence of the hybrid d orbitals (shown in Fig. 9.14) to the corresponding original d orbitals (shown in Fig. 9.13).

RADIAL DEPENDENCE OF p ORBITALS AND d ORBITALS

In the preceding discussions involving p and d orbitals we have been primarily concerned only with the angular part of the total wave function. However, the total probability density ψ^2 is the product of $|\Theta\Phi|^2$ multiplied by the radial function R^2. Curves of R^2 as a function of r for several p and d orbitals are given in Fig. 9.15. Note that the d orbitals show much less penetration than the p orbitals, which in turn show less penetration than s orbitals. *In systems containing more than one electron,* d electrons are held less tightly by the nucleus than are p electrons, which in turn are held less tightly than s electrons.

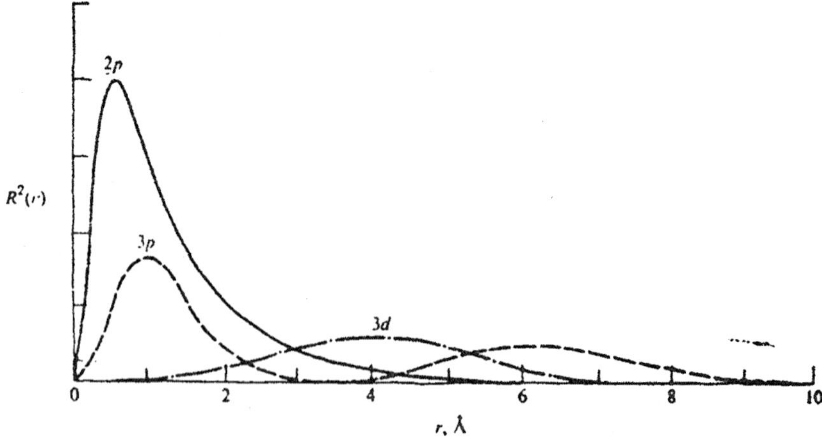

Fig. 9.15. The square $R^2(r)$ of the radial function for several states with angular dependence.

ELECTRON SPIN AND THE SPIN QUANTUM NUMBER

Although quantum mechanics was quite successful in accounting for atomic spectra, there still remained certain problems. For example, spectral lines for the alkali metals appear as closely spaced doublets rather than as the single lines predicted by nonrelativistic quantum-mechanical theory. Wolfgang Pauli attributed such spectral doublets to the existence of two closely spaced energy levels. In 1928 Dirac quantitatively accounted for the unexpected multiplicity of energy levels in terms of relativity.

According to relativity theory, the total energy E of a particle not only must include terms for kinetic and potential energy, it must also include a term for the relativitic rest mass energy m_0c^2. For example, the total energy of the particle in the one-dimensional box was given earlier as $E = n^2h^2/8mL^2$. The relativistic quantum-mechanical solution of the same problem requires that the total energy by very nearly expressed as

$$E = (n^2h^2/8mL^2) + m_0c^2, \tag{9.114}$$

where m_0 is the rest of the particle. However, since usual experimental methods measure only *differences* in energy levels, rather than energy levels themselves, the m_0c^2 term is not ordinarily observed experimentally.

In terms of non-relativistic treatment the operator for the z component of orbital angular momentum is

$$\hat{L}_z = \frac{h}{i2\pi} \cdot \frac{\partial}{\partial\varphi}.$$

In the *relativistic* treatment, Dirac showed that it was necessary to modify the operator for the z component of angular momentum to include a second term such that

$$\hat{J}_z = \frac{h}{i2\pi} \cdot \frac{\partial}{\partial\varphi} \pm \frac{1}{2}\frac{h}{2\pi}, \tag{9.115}$$

where $\hat{J}_z$ is the *relativistic* operator for the z component of angular momentum, J_z. Rewriting Eq. (9.115) we have

$$\hat{J}_z = \hat{L}_z \pm \frac{2}{2}\frac{h}{2\pi}, \tag{9.116}$$

The additional term, $\pm\frac{1}{2}h/2\pi$, in the operator provides for the existence of *two separate* allowed levels for the z component of angular momentum, instead of the single level which would be allowed if $\hat{J}_z$ where equal to $\hat{L}_z$ alone. One level is $\frac{1}{2}h/2\pi$ *above* the z component of angular momentum allowed by nonrelativistic quantum mechanics and the other one is $\frac{1}{2}h/2\pi$ *below* the level allowed by nonrelativistic quantum mechanics.

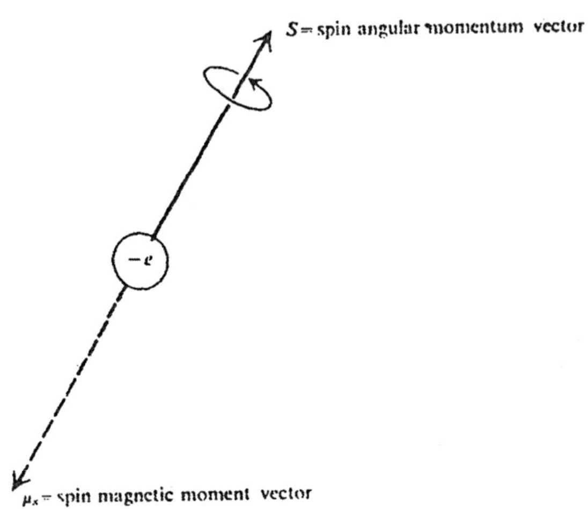

Fig. 9.16. Intrinsic spin of the electron on its own axis.

A classical interpretation of the two new levels may be given in terms of *intrinsic spin of the electron* on its own axis (although such an interpretation is not *required* by relativistic quantum mechanics). For the direction of spin shown in Fig. 9.16 the *spin angular momentum vector S* points upward (right- hand rule) and the spin magnetic moment vector μ_s points downward (left-hand rule). Just as the orbital angular momentum L is quantized according to the relationship

$$L = \sqrt{(l(l+1)}\,\frac{h}{2\pi},$$

where $l = 0, 1, 2, \ldots, n-1$, we can show that spin angular momentum S is given by

$$S = \sqrt{(s(s+1)}\,\frac{h}{2\pi}. \tag{9.117}$$

However, here the number s is restricted to the single value ½. That, is, in spinning on its own axis, the electron may have only one allowed value for the magnitude of its total spin angular momentum vector S.

We have earlier shown that the z component of orbital angular momentum, L_z, is given by

$$L_z = m \frac{h}{2\pi},$$

where $m = -l, \ldots, 0, \ldots, +l$. By comparison we can shown that the z component of spin angular momentum is given by

$$S_z = m_s \frac{h}{2\pi}, \tag{9.118}$$

where m_s, the *spin quantum number,* is allowed the values +½, −½. Thus, for given values of n, l, and m, only two values are allowed for S_z : one in which $m_s = +½$, in which case the spin angular momentum vector S points *upward* in an external magnetic filed directed upward, and one in which $m_s = -½$, in which case the spin angular momentum vector S points *downward* in an external magnetic field directed upward.

Thus, in the presence of an external magnetic field it is predicted that the spinning electron may assume either of two different closely spaced energy states. The Stern-Gerlach experiment confirmed this prediction by showing that if a beam of atoms of an alkali metal is passed between the poles of a strong magnet, the beam is split into two parts which may be collected separately. In the absence of an external magnetic field, for quantum states in which $l = 0$ (states of zero orbital angular momentum), the sates designated by $m_s = +½$ and $m_s = -½$ have identical energies and are therefore degenerate. For those states which have orbital angular momentum (for which l is greater than zero) the classical electron rotates on its own axis in the magnetic field produced by its own orbital motion. In this case, the S vector is allowed one of two different orientations with respect to the L vector. Each orientation has a slightly different energy level. This slight difference in energy levels between the $m_s = +½$ and $m_s = -½$ states leads to the appearance of closely spaced doublets in the atomic spectrum.

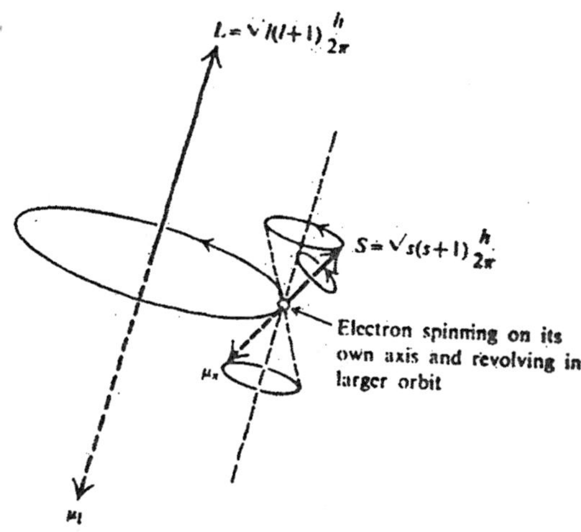

Fig. 9.17. An idealized classical picture of the relationship between the orbital angular momentum vector L and the spin angular momentum vector S, which is shown in the higher energy form of its two allowed positions. In the lower energy position, the S vector points downward at the same angle of precession.

Although relativity requires the introduction of a fourth quantum number (essentially due to the consideration of *time* as a fourth degree of freedom), we find it very convenient to adhere to the previous description for the state function ψ in terms of the quantum numbers n, l, and m and to account for electron spin by introducing a spin function α and β in the product of functions defining ψ, such that two total functions are possible. That is, for *spin up* ($m_z = +½$),

$$\psi_{nlmm_s} = R_{nl}\Theta_{lm}\Phi_m\alpha, \tag{9.119}$$

whereas, for *spin down* ($m_s = -½$),

$$\psi_{nlmm_s} = R_{nl}\Theta_{lm}\Phi_m\beta, \tag{9.120}$$

10

The Helium Atom

INTRODUCTION

The quantum-mechanical solutions considered so far have been direct. That is, we were able to solve directly the Schrodinger amplitude equations in order to evaluate wave functions and allowed stationary-state energies of systems. We were able to determine directly the ψ functions for two-particle systems such as the hydrogen atom and hydrogen-like ions. However, now in this part we will see that the nonrelativistic Schrodinger equation for the next simplest atom, the helium atom, which contains *three* particles, cannot be solved directly. Actually, the Schrodinger equation has not been solved directly for *any* system which contains more than two interacting particles. For all such systems we have to resort to approximation methods.

The helium atom, (Fig. 10.1) is a three-particle system which consists of two electrons and a nucleus whose mass is 4.0026 amu (6.6461 $\times$ 10^{-24} g), and whose charge is $+2e$. Let us assume that the total Schrodinger equation may be separated into two equations, one involving the translational energy of the atom and the other involving the energy of relative motion of the electrons and the nucleus. The translational energy associated with the motion of the center of mass of the atom in space will be ignored and we will concentrate our attention on the relative motion of the particles within the atom and on the energy of relative motion. In addition in our treatment of relative motion, we will *assume that the nucleus is stationary*. Although this is not exactly true, the mass of the nucleus is so much larger (about four thousand times larger) than the combined mass of the electrons that the resultant error is not significant. The total potential energy of the atom is the sum of the potential-energy terms for each of the possible two- particle interactions. That is, it is the sum of : (a) The potential energy of attraction between the first electron and the nucleus, $-2e^2/r_1$, where r_1 is the distance between the first electron and the nucleus. (b) The potential energy of attraction between the second electron and the nucleus, $-2e/r_2$, where r_2 is the distance between the second electron and the nucleus. (c) The potential energy of repulsion between the two electrons, $+e^2/r_{12}$, where r_{12} is the distance between the two electrons.

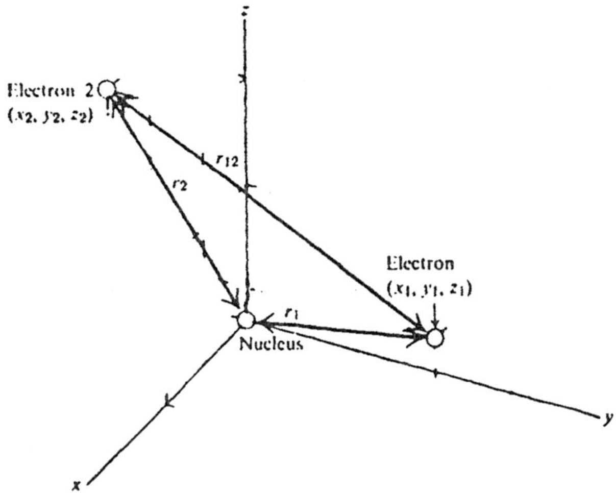

Fig. 10.1. The helium atom. The charge on the nucleus is +2*e* and the charge on each of the electrons is $-e$.

The total potential energy V for the atom may then be expressed as

$$V = -\frac{2e^2}{r_1} - \frac{2e^2}{r_2} + \frac{e^2}{r_{12}} \tag{10.1}$$

Since V is not dependent on time, the force field is conservative, and we may use the nonrelativistic time-independent Schrodinger amplitude equation

$$\hat{H}\psi = E\psi, \tag{10.2}$$

to calculate the family of wave functions of ψ and the corresponding energies of relative motion E. However, in writing Hamiltonian operator for the helium atom, we must take into account the sum of the kinetic energies of *two particles* (two electrons) and must, therefore, include a kinetic-energy operator term for *each* of the particles. Thus

$$\hat{H} = \frac{-h^2}{8\pi^2 m_e} \nabla_1^2 - \frac{h^2}{8\pi^2 m_e} \nabla_2^2 + V, \qquad (10.3)$$

where m_e is the mass of each electron. The first Laplacian operator, ∇_1^2, operates only on the coördinates (x_1, y_1, z_1) of the first electron, and the second Laplacian operator, ∇_2^2, operates only on the coordinates (x_2, y_2, z_2) of the second electron that is,

$$\nabla_1^2 = \frac{\partial^2}{\partial x_1^2} + \frac{\partial^2}{\partial y_1^2} + \frac{\partial^2}{\partial z_1^2} \qquad (10.4)$$

and

$$\nabla_2^2 = \frac{\partial^2}{\partial x_2^2} + \frac{\partial^2}{\partial y_2^2} + \frac{\partial^2}{\partial z_2^2} \qquad (10.5)$$

on substituting Eqs. (10.1) and (10.3) into Eq. (10.2) we get the complete Schrodinger amplitude equation for relative motion in the helium atom :

$$\frac{-h^2}{8\pi^2 m_e} (\nabla_1^2 \psi + \nabla_2^2 \psi) - \frac{2e^2\psi}{r_1} - \frac{2e^2\psi}{r_2} - \frac{e^2\psi}{r_{12}} = E\psi. \qquad (10.6)$$

Eq. (10.6) has never been solved by any direct method; we are forced to rely on certain approximation methods, to calculate the ground-state energy of the helium atom.

GROUND-STATE ENERGY OF THE HELIUM ATOM

First Approximation Ignore Electron Repulsion

The presence of the electron repulsion term in Eq. (10.6) prevents us from obtaining a direct solution. As a first approximation, therefore, we assume that the electrons do not repel one another and proceed with the calculation. If the electron repulsion energy is ignored, the total potential energy of the free helium atom is given by the relation

$$V = -\frac{2e^2}{r_1} - \frac{2e^2}{r_2}, \tag{10.7}$$

and the Schrodinger amplitude equation is

$$\left[\frac{-h^2}{8\pi^2 m_e}(\nabla_1^2 + \nabla_2^2) - 2e^2{}_{r_1} - \frac{2e^2}{r_2}\right]\psi^0 = E^0\psi^0, \tag{10.8}$$

where the eigenvalues E^0 are the *stationary-state energies of relative motion of the helium atom assuming no electron repulsion* and the eigenfunctions ψ^0 are the *wave functions for the helium atom in which electron repulsion is ignored.* The associated nonrepulsion Hamiltonian operator $\hat{H}^0$ is then

$$\hat{H}^0 = \frac{-h^2}{8\pi^2 m_e}(\nabla_1^2 + \nabla_2^2) - \frac{2e^2}{r_1} - \frac{2e^2}{r_2}, \tag{10.9}$$

so that we can write Eq. (10.8) as

$$\hat{H}^0\psi^0 = E^0\psi^0. \tag{10.10}$$

In order to separate the variables associated with the first electron from those associated with the second electron, let us assume that the total wave function ψ^0 may be expressed as the product of two functions, such that the first function, φ_1, depends only on the coordinates of the first electron and the second function, φ_2, depends only on the coordinates of the second electron. That is,

$$\psi^0(x_1, y_1, z_1, x_2, y_2, z_2) = \varphi_1(x_1, y_1, z_1)\,\varphi_2(x_2, y_2, z_2), \tag{10.11}$$

or simply as

$$\psi^0 = \varphi_1\varphi_2. \tag{10.12}$$

On substituting Eq. (10.12) into Eq. (10.8), followed by division of both sides by $\varphi_1\varphi_2$, we get

$$\left[-\frac{1}{\varphi_1}\frac{h^2}{8\pi^2 m_e}\nabla_1^2\varphi_1 - \frac{2e^2}{r_1}\right] + \left[-\frac{1}{\varphi_2}\frac{h^2}{8\pi^2 m_e}\nabla_2^2\varphi_2 - \frac{2e^2}{r_2}\right] = E^0 \tag{10.13}$$

But the value of the first term in brackets in Eq. (10.13) depends only on the coordinates of the first electron and is independent of the coordinates of the second electron, and the value of the second term in Eq. (10.13) depends only on the coordinates of the second electron and is independent of the coordinates of the first electron. Since the sum of both terms is a constant and, since the two terms are independent on one another, each term must separately be equal to a constant, that is,

$$-\frac{1}{\varphi_1}\frac{h^2}{8\pi^2 m_e}\nabla_1^2\varphi_1 - \frac{2e^2}{r_1} = E_1, \tag{10.14}$$

and

$$-\frac{1}{\varphi_2}\frac{h^2}{8\pi^2 m_e}\nabla_2^2\varphi_2 - \frac{2e^2}{r_2} = E_2, \tag{10.15}$$

where

$$E^0 = E_1 + E_2. \tag{10.16}$$

Equations (10.14) and (10.15) may each be rearranged to the more familiar forms of the Schrodinger amplitude equations :

$$-\frac{h^2}{8\pi^2 m_e}\nabla_1^2\varphi_1 - \frac{2e^2\varphi_1}{r_1} = E_1\varphi_1, \tag{10.17}$$

and

$$-\frac{h^2}{8\pi^2 m_e}\nabla_2^2\varphi_2 - \frac{2e^2\varphi_2}{r_2} = E_2\varphi_2, \tag{10.18}$$

Equations (10.17) and (10.18) are *each* recognized as being identical in form to Eq. (9.1) for the hydrogen-like ion, where $Z = 2$, that is, where the ion is He^+.

In Chapter 9 we solved an equation identical in form to Eqs. (10.17) and (10.18), and the resultant stationary-state energies of relative motion are given by Eq. (9.68). Through identical solution, where now m_e is substituted for μ, we have

$$E_1 = -2\pi^2 m_e Z^2 e^4/n_1^2 h^2, \tag{10.19}$$

and

$$E_2 = -2\pi^2 m_e Z^2 e^4/n_2^2 h^2, \tag{10.20}$$

But we have shown earlier that E_H, the energy of the hydrogen atom in the *ground state*, is given (Eq. 9.69) as $-2\pi^2\mu e^4/h^2$. We'll now assume that within the accuracy of our present approximation, μ may be replaced by m_e without significant error, so that

$$E_H = -2\pi^2 m_e e^4/h^2, \tag{10.21}$$

and since Z = 2 for the He^+ ion, E_1 and E_2, as given by Eqs. (10.19) and (10.20), may be expressed as

$$E_1 = 4E_H/n_1^2 \text{ and } E_2 = 4E_H/n_2^2,$$

so that E^0, from Eq. (10.16), becomes

$$E^0 = 4E_H\left(\frac{1}{n_1^2} + \frac{1}{n_2^2}\right) \tag{10.22}$$

If each of the electrons in the nonrepulsion model of the helium atom is in its ground state, that is, if $n_1 = 1$ and $n_2 = 1$, the ground-state energy of the helium atom (assuming no electron repulsion) is

$$E^0 = 2E_{HE^+} = 8E_H, \tag{10.23}$$

where E_{He^+} is the *ground-state* energy of the helium *ion*. But E_H was shown to be -2.18×10^{-19} J, which is equivalent to -13.6 eV. Thus E^0, the ground-state energy of the helium *atom* in our first crude approximation is

$$E^0 = 9(-13.6 \text{ eV}) = -108.8 \text{ eV}. \tag{10.24}$$

The best *experimental* value for the energy E_{He} of the helium atom in the ground state is given in terms of the *first ionization potential I*, which is the *minimum* energy absorbed in the reaction in which the first electron leaves the ground state of the helium atom to become a free electron :

$$\text{He (ground state)} \rightarrow \text{He}^+ + \text{e}. \tag{10.25}$$

Since the minimum energy corresponds to an ejected electron having no translation kinetic energy, we can write the energy balance for Eq. (10.25) as

$$E_{He} + I = E_{He^+}. \tag{10.26}$$

But we are able to solve the Schrodinger amplitude equation for E_{He^+} directly and have shown (Eq. 10.23) that

$$E_{He^+} = E^0/2 = -54.5 \text{ eV},$$

and I is experimentally determined to be 24.6 eV. Thus, from Eq. (10.26), the experimental value for the ground-state energy of the helium atom is given as

$$E_{\text{He}} = -54.5 - 24.6 = -79.0 \text{ eV}. \qquad (10.27)$$

A comparison of E_{He} (experimental) from Eq. (11.27) with E^0 (calculated) in Eq. (10.24) indicates that in our first approximation method, the calculated value of the ground-state energy of the helium atom is 38% lower than the experimental value. Such a large error indicates that electron repulsion cannot be ignored.

Second Approximation (A First Order Perturbation Method)

We have indicated that the correct Schrodinger wave equation for the helium atom is

$$\hat{H}\psi = E\psi, \qquad (10.2)$$

where

$$\hat{H} = \frac{-h^2}{8\pi^2 m_e}\nabla_1^2 - \frac{h^2}{8\pi^2 m_e}\nabla_2^2 - \frac{2e^2}{r_1} - \frac{2e^2}{r_2} + \frac{e^2}{r_{12}}. \qquad (10.28)$$

Equation (10.2) cannot be solved directly, which means that we cannot directly evaluate the *true wave functions,* ψ. We have shown, however, that if we ignore electron repulsion, we can solve *directly* the equation

$$\hat{H}^0\psi^0 = E^0\psi^0, \qquad (10.29)$$

where, from Eq. (10.9),

$$\hat{H}^0 = \frac{-h^2}{8\pi^2 m_e}\nabla_1^2 - \frac{h^2}{8\pi^2 m_e}\nabla_2^2 - \frac{2e^2}{r_1} - \frac{2e^2}{r_2}, \qquad (10.30)$$

and where the eigenfunctions ψ^0 are the *wave functions for the helium atom in which electron repulsion is ignored.* Substitution of Eq. (10.30) into Eq. (10.28) gives

$$\hat{H} = \hat{H}^0 + \frac{e^2}{r_{12}}, \qquad (10.31)$$

substituting Eq. (10.31) into (10.2) we have

$$\left(\hat{H}^0 + \frac{e^2}{r_{12}}\right)\psi = E\psi, \qquad (10.32)$$

an equation which cannot be directly solved. Let's now compare Eqs. (10.29) and (10.32). The difference in the operators in the two eigenvalue equations is the repulsion term, e^2/r_{12}. The smaller the value of e^2/r_{12}, the more nearly identical will the two equations be. We have seen, in fact, that if we let e^2/r_{12} equal zero (ignore electron repulsion), the two equations become identical. That is, E values are then given by E^0 values, and ψ functions are given by ψ^0 functions.

In using the *perturbation method* we'll assume that e^2/r_{12} is small enough to be considered as a minor modification or *perturbation* of the operator $\hat{H}^0$, an operator for which we may directly calculate the eigenfunctions ψ^0. We further hope that the perturbation will be small enough so that for a given state ψ^0 will not be too much different from ψ (which we cannot directly evaluate). Thus, in evaluating the ground-state energy of the helium atom in a gives state, we'll combine the use of the correct Hamiltonian operator $\hat{H} = \hat{H}^0 + e^2/r_{12}$ with an incorrect wave function ψ^0 for that state which is regarded to be fairly close to what the correct wave function would be. The nonrepulsion operator $\hat{H}^0$ is said to be perturbed to the first order by the term e^2/r_{12}, and the present method of approximation is called a *first-order* perturbation method. The previous approximation method in which we ignored electron repulsion altogether is called a *zero- order* perturbation method, since we used $\hat{H}^0$ directly as the operator. That is, we added *no* perturbation terms to the nonrepulsion Hamiltonian that we used. It is important to note that for a give state ψ^0 is *not* an eigenfunction of $\hat{H}$ and therefore the operation of $\hat{H}$ on ψ^0 will not yield an exact energy eigenvalue. Rather, because of the perturbation of energy due to the repulsion term, the value of E obtained by the operation of $\hat{H}$ on ψ^0 for a given state will depend on the relative positions of the two electrons so that an *approximate* average value of energy, E, must be calculated through use of the mean-value postulate as

$$E = \frac{\int_{-\infty}^{\infty} \psi^{0*} \hat{H} \psi^0 \, d\tau}{\int_{-\infty}^{\infty} \psi^{0*} \psi^0 \, d\tau} = \frac{\int_{-\infty}^{\infty} \psi^{0*} (\hat{H}^0 \psi^0 + e^2 \psi^0 / r_{12}) \, d\tau}{\int_{-\infty}^{\infty} \psi^{0*} \psi^0 \, d\tau}. \quad (10.33)$$

Substitution of Eq. (10.29) into Eq. (10.33) gives

$$E = \frac{\int_{-\infty}^{\infty} \psi^0 E^0 \psi^{0*} \, d\tau + \int_{-\infty}^{\infty} \psi^{0*} (e^2/r_{12}) \psi^0 \, d\tau}{\int_{-\infty}^{\infty} \psi^{0*} \psi^0 \, d\tau}, \quad (10.34)$$

but, since for any given state E^0 is constant,

$$E = E^0 + \frac{\int_{-\infty}^{\infty} \psi^{0^*} (e^2/r_{12}) \psi^0 \, d\tau}{\int_{-\infty}^{\infty} \psi^{0^*} \psi^0 \, d\tau}. \tag{10.35}$$

According to Eq. (10.35), the approximate *first-order* energy E for the helium atom in a given state is given by the sum of E^0, the energy of the *nonperturbed* atom in that same state in which electron repulsion is ignored (the *zero-order energy*), and a term which is equivalent to the approximate average potential energy of electron repulsion over all space. Note that the second term on the right-hand side of Eq. (10.35) would be, according to the mean-value postulate, the mean value of e^2/r_{12} if ψ^0 were used as the correct wave function, since the quantum-mechanical operator for e^2/r_{12} is also e^2/r_{12}. The second term on the right-hand side of Eq. (10.35) is called the *perturbation energy* E'.

We may therefore write

$$E = E^0 + E', \tag{10.36}$$

where the *perturbation energy* is

$$E' = \frac{\int_{-\infty}^{\infty} \psi^{0^*} (e^2/r_{12}) \psi^0 \, d\tau}{\int_{-\infty}^{\infty} \psi^{0^*} \psi^0 \, d\tau}. \tag{10.37}$$

Thus, in using the perturbation method we think of the energy E of a given state of the helium atom as consisting of the total of E^0, the exact sum of the kinetic energy and the potential energy of attraction in that same state calculated by ignoring repulsion, and E', the approximate mean potential energy of repulsion, which in this case is a perturbation energy.

Since we wish to evaluate E' for the *ground state* of the helium atom ($n_1 = 1$ and $n_2 = 1$), we must use 1s orbitals for each of the electrons. Thus, from Table 9.1, where $Z = 2$,

$$\varphi_1 = \varphi_1^* = \left(\frac{1}{\pi}\right)^{1/2} \left(\frac{2}{a_0}\right)^{3/2} e^{-2r_1/a_0}, \tag{10.38}$$

and

$$\varphi_2 = \varphi_2^* = \left(\frac{1}{\pi}\right)^{1/2}\left(\frac{2}{a_0}\right)^{3/2} e^{-2r_2/a_0}, \tag{10.39}$$

since $\psi^0 = \varphi_1\varphi_2$ and $\psi^{0^*} = \varphi_1^*\varphi_2^*$,

$$\psi^0 = \psi^{0^*} = \varphi_1\varphi_2 = \left(\frac{1}{\pi}\right)\left(\frac{2}{a_0}\right)^3 e^{-2r_1/a_0}\, e^{-2r_2/a_0}. \tag{10.40}$$

But φ_1 and φ_2, as given by Table 9.1, are normalized, so that

$$\int_{-\infty}^{\infty} \varphi_1^*\varphi_1\, d\tau_1 = 1,$$

$$\int_{-\infty}^{\infty} \varphi_2^*\varphi_2\, d\tau_2 = 1,$$

and

$$\int_{-\infty}^{\infty} \psi^{0^*}\psi^0\, d\tau = \int_{-\infty}^{\infty}\int_{\infty}^{\infty} \varphi_1^*\varphi_1 \cdot \varphi_2^*\varphi_2\, d\tau_1\, d\tau_2 = 1. \tag{10.41}$$

Substitution of Eqs. (10.40) and (10.41) into Eq. (10.37) yields

$$E' = \left(\frac{8}{\pi a_0^3}\right)^2 \int_{-\infty}^{\infty} \frac{e^2}{r_{12}} (e^{-2r_1/a_0}e^{-2r_2/a_0})^2\, d\tau \tag{10.42}$$

or, in terms of the coordinates of each of the particles,

$$E' = \left(\frac{8}{\pi a_0^3}\right)^2 \int_{-\infty}^{\infty}\int_{-\infty}^{\infty} \frac{e^2}{r_{12}} (e^{-2r_1/a_0}e^{-2r_2/a_0})^2\, d\tau_1\, d\tau_2. \tag{10.43}$$

The integration of Eq. (10.43) is algebraically difficult. The final result for the perturbation energy is

$$E' = -\tfrac{5}{2}E_H = -\tfrac{5}{2}(-13.6\text{eV}) = +34.0\text{eV}. \tag{10.44}$$

Substitution of the value for E^0 from Eq. (10.24) and E′ from Eq. (10.44) into Eq. (10.36) gives

$$E = -108.8 + 34.0 = -74.8\text{ eV} \tag{10.45}$$

for the approximate ground-state energy of the helium atom, as calculated by a first-order perturbation method. A comparison of E (−74.8 eV) with the experimental value (−79.0 eV) shows that the calculated value is 5.3% too high (algebraically). If we use the calculated E value to compute a value for the ionization potential, we obtain

$$I_{calc} = E_{He^+} - E$$

$$= -54.4 - (-74.8) = 20.4 \text{eV}.$$

As compared with the experimental value of I, which is 24.6 eV, the value calculated by the perturbation method is 17% too low. It thus appears that a comparison of calculated and experimental first ionization potentials is a more sensitive measure of the accuracy of a given approximation method than is a comparison of ground-state energies, the reason being that the ionization potential is a measure of a *difference* in energy levels.

The accuracy of the perturbation method is dependent primarily on the accuracy with which ψ^0 reflects the form of the correct wave function, ψ. For small perturbations on $\hat{H}$, ψ is apparently close to ψ^0.

The Variation Theorem

In using the perturbation method we made use of the mean-value postulate in order to estimate the ground-state energy of the helium atom. However, the energy which we obtained was not really equal to the quantum-mechanical expectation value because ψ^0 for the ground state was only an approximation of the true wave function. A better choice of trial wave function (better than ψ^0) would have given an energy value nearer to the true expectation value. The *variation method*, provides a systematic approach for trying and comparing different guesses at the form of the correct wave function. The *variation theorem* is stated as follows :

If we use any well-behaved approximate function $\tilde{\psi}$ in evaluating the integral in the mean-value postulate, the value of the expectation energy $\tilde{E}$ so obtained will always be algebraically equal to or greater than the true ground-state energy E_0 of the system. That is,

$$\tilde{E} = \frac{\int_{-\infty}^{\infty} \tilde{\psi}^* \hat{H} \tilde{\psi}\, d\tau}{\int_{-\infty}^{\infty} \tilde{\psi}^* \tilde{\psi}\, d\tau} \geq E_0, \qquad (10.46)$$

where $\hat{H}$ is the complete and correct Hamiltonian operator for the system of interest.

It is important to note that E in Eq. (10.46) is *not* the true expectation energy. It is merely a trial energy evaluated from the same expression which yields the true expectation energy *when the true wave functions is used.* One way of using the variation theorem would be to keep trying well-behaved $\tilde{\psi}$ functions until we are satisfied that no appreciably lower energy can be obtained, at which point the $\tilde{\psi}$ function would be very close to ψ^0, the wave function for the true ground state. The mathematics is quite burdensome, and using high-speed computers in such a trial and error approach is desirable.

Let us consider a very simple problem in order to illustrate the method. Let us assume that we are not able to calculate directly the ground-state energy for a particle of mass m in a one- dimensional box of length L. That is, assume that we cannot directly solve the Schrodinger equation

$$\hat{H}\psi = \frac{-h^2}{8\pi^2 m}\frac{d^2\psi}{dx^2} = E\psi. \tag{10.47}$$

Instead, we'll devise a trial function, $\tilde{\psi}$, which is well-behaved one obeys the boundary conditions, and insert the trial function into Eq. (10.46). One such acceptable trial function which goes to zero at $x =$ 0 and at $x = L$ is given by

$$\tilde{\psi} = x(L - x) = Lx - x^2. \tag{10.48}$$

Insertion of the correct form for $\hat{H}$ and the trial form for $\tilde{\psi}$ into Eq. (10.46) yields

$$E = \frac{\int_0^L (Lx - x^2)\left(\frac{-h^2}{8\pi^2 m}\frac{d^2}{dx^2}\right)(Lx - x^2)\,dx}{\int_0^L (Lx - x^2)(Lx - x^2)\,dx}, \tag{10.49}$$

which, after simple differentiation and integration, becomes

$$E = 1.013\,\frac{h^2}{8mL^2}. \tag{10.50}$$

We know that the ground-state energy E_0 for the particle in the one-dimensional box was given as

$$E_0 = h^2/8mL^2, \tag{10.51}$$

so that Eq. (10.50) may be written as

$$\bar{E} = 1.013\, E_0.$$

Thus, in the above particularly simple case, the expectation energy calculated by the variation method is surprisingly close to and slightly larger than the true ground-state energy. The enexpected accuracy of our first approximate calculation results from the fortunate choice of a trial function which is really not too far different from the true function for the ground state. The variation theorem states, however, that *any* trial function will yield a value of $\bar{E}$ which is equal to or larger than E_0.

THIRD APPROXIMATION

A Variation Method

Let us now use the variation theorem to approximate the ground-state energy of the helium atom. As a first step, we must choose a good trial function for helium. The unperturbed function for the ground state of helium was shown to be

$$\psi^0 = \varphi_1\varphi_2 = \left(\frac{1}{\pi}\right)\left(\frac{2}{a_0}\right)^3 e^{-2r_1/a_0}\, e^{-2r_2/a_0}, \qquad (10.52)$$

and we'll guess that the correct eigenfunction is not too different in form. Eq. (10.46) is identical to Eq. (10.33) if ψ^0 is substituted for $\tilde{\psi}$. In order to construct a trial function more accurate than ψ^0, we note that the negative charge cloud of the first electron results in an effective screening of the positive nuclear charge from the second electron. That is, because of the presence of the first electron, the second electron in effect "sees" a nucleus whose positive charge is somewhat less than $+2e$. The effect of the second electron on the first electron is considered to be the same. Therefore, a promising trial modification of Eq. (10.52) would result from the reduction of the nuclear charge from $+2e$ to an *effective value* of $+Z'e$ so that the trial function is written, through substitution of Z' for 2 in Eq. (10.52), as

$$\tilde{\psi} = \left(\frac{1}{\pi}\right)\left(\frac{Z'}{a_0}\right)^3 e^{(-Z'r_1/a_0}\, e^{-Z'r_2/a_0}, \qquad (10.53)$$

We'll now proceed to leave Z' undetermined and insert $\tilde{\psi}$ from Eq. (10.53) into Eq. (10.46) in order to obtain a value for $\bar{E}$ in terms of Z'. We'll then select the value of Z' which minimizes $\bar{E}$, that is, which brings $\bar{E}$ as close as possible to the true ground- state energy. Since the $\tilde{\psi}$ function given by Eq. (10.53) may be varied by choice of value for Z', it is called a *variation function*. Because the variation function is not complex and is already normalized, Eq. (10.46) may be written as

$$E = \int_{-\infty}^{\infty} \tilde{\psi} \hat{H} \tilde{\psi} \, d\tau. \tag{10.54}$$

Substitution of Eq. (10.53) into Eq. (10.54) yields

$$E = \left(\frac{1}{\pi}\right)^2 \left(\frac{Z'}{a_0}\right)^6 \int_{-\infty}^{\infty} \int_{-\infty}^{\infty} (e^{-Z'r_1/a_0} e^{-Z'r_2/a_0}) \hat{H} (e^{-Z'r_1/a_0} e^{-Z'r_2/a_0}) \, d\tau_1 \, d\tau_2, \tag{10.55}$$

where

$$\hat{H} = \frac{-h^2}{8\pi^2 m_e} (\nabla_1^2 + \nabla_2^2) - \frac{2e^2}{r_1} - \frac{2e^2}{r_2} + \frac{e^2}{r_{12}}. \tag{10.56}$$

The correct solution of Eq. (10.55) yields

$$E = [-2(Z')^2 + 27/4 Z'] \left[\frac{-2\pi^2 m_e e^4}{h^2}\right],$$

or, through substitution of Eq. (10.21)

$$E = [-2(Z')^2 + 27/4 Z'] \, E_H \tag{10.57}$$

The best value for Z' will be that value for which E is minimum, that is, that value of Z' for which $dE/dZ' = 0$. The first derivative of Eq. (10.57) equated to zero is

$$\frac{dE}{dZ'} = (-4Z' + 27/4) E_H = 0,$$

from which

$$Z' = 27/16.$$

We may now calculate E by inserting the value $27/16$ for Z' in Eq. (10.57) :

$$E = 5.70 E_H = 5.70(-13.6 \text{ eV}) = -77.5 \text{ eV}. \tag{10.58}$$

The experimental value for the ground-state energy of the helium atom is -79.0 eV. In comparison, then, the value -77.5 eV yielded by the variation theorem method is 1.9%. The ionization potential calculated by the variation method is

$$I = E_{He^+} - \bar{E} = -54.4 - (-77.5) = 23.1 \text{ eV}.$$

Since the experimental value of I is 24.6 eV, the calculated result is 6.1% less.

Of the three approximation methods we considered so far, the variation method provides calculated values for the ground-state energy and ionization potential in best agreement with experimental values. It is possible to further improve the results of the variation method if we use a trial function which takes into account the tendency of the two electrons to avoid one another. That is, in addition to having mutual nuclear screening effects on one another, the two like-charged electrons in the helium atom correlate their movements and positions in such a way as to remain as far apart as possible, and the trial function should take such *electron correlation* into account. We can introduce a second parameter c into the trial function given by Eq. (10.53) such that

$$\tilde{\psi} = \left(\frac{1}{\pi}\right)\left(\frac{Z'}{a_0}\right)^3 (1 + cr_{12})e^{-Z'r_1/a_0}e^{-Z'r_2/a_0}. \qquad (10.59)$$

The incorporation of the additional polynomial term $(1 + cr_{12})$ causes $\tilde{\psi}$ to be larger when r_{12} is larger (if $c > 0$). Thus the probability $\tilde{\psi}^2$ for a particular distribution will also be greater as r_{12} increases. That is, distributions in which the electrons are farther apart become more probable. Using Eq. (10.59) in the variation method, we are able to calculate a function for $\bar{E}$ in terms of c and Z'. Minimization of $\bar{E}$ in terms of both c and Z' then leads to a ground-state energy for the helium atom which is less than 0.4 eV larger than the experimental value.

Excited States

Let us now consider the application of a first-order perturbation method to the calculation of the energy of an excited state of helium in which one of the electrons is promoted from the 1s orbital to some higher-energy orbital, for example, the 2s or 2p orbital. We will initially ignore the effects of electron repulsion i.e., we will initially consider the unperturbed system. A satisfactory solution for the unperturbed Schrodinger equation (Eq. 10.8) may once again be written as a product,

$$\psi_{1,2} = \psi_A(1)\psi_B(2), \tag{10.60}$$

where $\psi_A(1)$ signifies that electron 1 is in the ψ_A orbital and $\psi_B(2)$ signifies that electron 2 is in the ψ_B orbital. For example, ψ_A may be the 1s orbital and ψ_B may be the 2s orbital. If E_A is the energy of the first electron in the ψ_A orbital and E_B is the energy of the second electron in the ψ_B orbital, then the total zero-order energy of the system ignoring repulsion would be $E_A + E_B$.

However, it is apparent that the function

$$\psi_{2,1} = \psi_A(2)\psi_B(1) \tag{10.61}$$

is an *equally satisfactory* solution to the unperturbed Schrodinger equation. In writing $\psi_{1,2}$ we assume that electron 2 is in the ψ_A orbital and electron 1 is in the ψ_B orbital. But it is experimentally impossible to distinguish between the two different distributions represented by $\psi_{1,2}$ and $\psi_{2,1}$. That is, *we cannot determine which electron is in which orbital.*

So far we have been concerned only with ensuring that the wave function for a given system satisfy the Schrodinger equation. However, *if the wave function is to completely define a system containing two electrons, it must also account for the indistinguishability of the electrons.* In order for a wave function ψ for a two-electron system to account satisfactorily for electron indistinguishability, it is necessary that the resultant electron probability density ψ^2 remain the same when the coordinates of the two electrons are interchanged, since either of the two indistinguishable configurations must lead to the same electron probability distribution. This means that the interchange of electron coordinates between the two orbitals, must yield either the original function ψ or the negative of the original function $(-\psi)$ since the square of either is ψ^2.

We will now determine whether $\psi_{1,2}$ as given by Eq. (10.60) is a satisfactory function for the system in terms of accounting for electron indistinguishability. The interchange of electron coordinates (if ψ_A and ψ_B were spherically symmetrical, this would involve exchanging r_1 and r_2 between the two orbitals) must yield either $\psi_{1,2}$ or $(-\psi_{1,2})$. Actually, the interchange yields

$$\psi_A(2)\psi_B(1),$$

which is equal to neither $\psi_{1,2}$ nor $(-\psi_{1,2})$. Thus $\psi_{1,2}$ as given by Eq. (10.60) is *not* a satisfactory representation of the state in which the electrons are indistinguishable and each is in a different orbital. By the same argument, neither is $\psi_{2,1}$ as given by Eq. (10.61) satisfactory.

Note that if the two electrons were in the *same* orbital, as in the ground-state calculation for the helium atom, the wave function

$$\psi_{2,1} = \psi_{1s}(2)\psi_{1s}(1)$$

is identical to the interchanged function

$$\psi_{1,2} = \psi_{1s}(1)\psi_{1s}(2)$$

By virtue of their being in the same orbital, the two electrons in the ground state are automatically indistinguishable in terms of space coordinates.

Returning again to the excited state; even though neither the $\psi_{1,2}$ function as given by Eq. (10.60) nor the $\psi_{2,1}$ function as given by Eq. (10.61) is in itself a satisfactory function in terms of electron indistinguishability, we may create two new satisfactory functions linear combination of Eqs. (10.60) and (10.61). Because the two states represented by $\psi_{1,2}$ and $\psi_{2,1}$ correspond to the same eigenvalue energy, any linear combination of $\psi_{1,2}$ and $\psi_{2,1}$ is a satisfactory solution to the unperturbed Schrodinger equation. The two linear combinations which are *also* satisfactory with respect to the criterion of *electron indistinguishability* are the sum and the difference of Eq. (10.60) and Eq. (10.61) :

$$\psi_s = (1/\sqrt{2})\,[\psi_A(1)\psi_B(2) + \psi_A(2)\psi_B(1)], \qquad (10.62)$$

and

$$\psi_a = (1/\sqrt{2})\,[\psi_A(1)\psi_B(2) + \psi_A(2)\psi_B(1)], \qquad (10.63)$$

The coefficient, $1/\sqrt{2}$, is required to ensure that ψ_s and ψ_a are normalized. The interchange of electron coordinates in ψ_s yields

$$(1/\sqrt{2})\,[\psi_A(2)\psi_B(1) + \psi_A(1)\psi_B(2)] = \psi_s,$$

which is identical to the original function. The interchange of electron coordinates in ψ_a yields

$$(1/\sqrt{2})\,[\psi_A(2)\psi_B(1) + \psi_A(1)\psi_B(2)] = -\psi_a,$$

which is the negative of the original function. Because the sign of ψ_s does *not* change when the coordinates of the electrons are interchanged, it is said to be *symmetric*; hence the subscript s. Because the sign of ψ_a changes when the coordinates of the electrons are interchanged, it is said to be *antisymmetric,* hence the subscript a.

Either of the function ψ_s or ψ_a is a satisfactory wave equation for the unperturbed two-electron system in which one electron is in the ψ_A orbital and the other electron is in the ψ_B orbital because : (a) Either ψ_s or ψ_a is a satisfactory solution to the unperturbed Schrodinger equation and, (b) Either ψ_s or ψ_a satisfactorily accounts for the indistinguishability of electrons.

In the unperturbed state, ψ_s and ψ_a correspond to degenerate states; however the degeneracy will be seen to disappear when we perturb the system by bringing electron repulsion into play.

ELECTRON SPINS

If an atom contains two electrons, one with a spin momentum component of $+\frac{1}{2}h/2\pi$ (spin *up*) with respect to a specified z-axis and the other with a spin momentum component of $-\frac{1}{2}h/2\pi$ (spin *down*) with respect to a specified z-axis, it is possible experimentally to determine only that the *net spin angular momentum is zero* with respect to the specified axis. *We cannot determine which of the two electrons has which spin.* In order that the wave function ψ might also satisfactorily account for the indistinguishability of electrons with respect to their intrinsic spins, it is necessary that we be able to interchange the spins of any pair of electrons without changing the value of ψ^2.

The total wave function, including spin, may be expressed as the product of the spatial function $R\Theta\Phi$ times either of two spin functions α or β (chapter 9 Eqs. 9.119 and 9.120). We arbitrarily called α the function for *spin up,* in which $m_s = +\frac{1}{2}$, and β the function for *spin down,* in which $m_s = -\frac{1}{2}$. Thus, for an electron in the ψ_A orbital, the total wave function including electron spin may be either $\psi_A\alpha$ or $\psi_A\beta$, and for an electron in the ψ_B orbital, the total wave function may be either $\psi_B\alpha$ or $\psi_B\beta$.

If for an unperturbed *two-electron* system we assume that the interaction between spin and orbital motion is negligible, we may consider the complete wave function for the system to be the product of a suitable spatial function for the two electrons times a suitable spin function for the two electrons. It has already shown that when one electron is in the ψ_A hydrogen-like space orbital and the other electron is in the ψ_B hydrogen-like space orbital, the only two-electron spatial functions which are satisfactory are given by the symmetric function ψ_s (Eq. 10.62), and the antisymmetric function ψ_a (Eq. 10.63). By a similar argument concerning spin indistinguishability, in order for the spin function to be a suitable representation for the pair of electrons, it must either remain the same or at the most change signs when the spins of the two electrons are interchanged. It is easily shown that the only spin functions for a pair of electrons which satisfactorily account for the indistinguishability of intrinsic spins are

$$\alpha(1)\alpha(2) \qquad \text{(symmetric)}, \qquad (10.64)$$

$$\beta(1)\beta(2) \qquad \text{(symmetric)}, \qquad (10.65)$$

$$(1/\sqrt{2})[\alpha(1)\beta(2)+\alpha(2)\beta(1)] \qquad \text{(symmetric)}, \qquad (10.66)$$

$$(1/\sqrt{2})[\alpha(1)\beta(2)-\alpha(2)\beta(1)] \qquad \text{(antisymmetric)}, \qquad (10.67)$$

where $1/\sqrt{2}$ is once more used as a normalized factor.

Interchange of electrons in the first three of the above functions reproduces the original function. That is, they are symmetric spin functions. The interchange of electrons in Eq. (10.67), however, reverses the sign of the spin function. Hence, the last spin function is an antisymmetric spin function.

There are eight complete wave functions which satisfy the unperturbed Schrodinger equation and which *also* satisfy the criterion of indistinguishability of electron coordinates *and* electron spins they are:

$$\frac{1}{\sqrt{2}}[\psi_A(1)\psi_B(2)+\psi_A(2)\psi_B(1)][\alpha(1)\alpha(2)] \qquad \text{(symmetric)}, \qquad (10.68)$$

$$\frac{1}{\sqrt{2}}[\psi_A(1)\psi_B(2)+\psi_A(2)\psi_B(1)][\beta(1)\beta(2)] \qquad \text{(symmetric)}, \qquad (10.69)$$

$$\frac{1}{\sqrt{2}}[\psi_A(1)\psi_B(2)+\psi_A(2)\psi_B(1)]\frac{1}{\sqrt{2}}[\alpha(1)\beta(2)+\alpha(2)\beta(1)]$$

$$\text{(symmetric)}, \qquad (10.70)$$

$$\frac{1}{\sqrt{2}}[\psi_A(1)\psi_B(2) + \psi_A(2)\psi_B(1)]\frac{1}{\sqrt{2}}[\alpha(1)\beta(2) - \alpha(2)\beta(1)]$$
(antisymmetric), (10.71)

$$\frac{1}{\sqrt{2}}[\psi_A(1)\psi_B(2) - \psi_A(2)\psi_B(1)][\alpha(1)\alpha(2)] \quad \text{(antisymmetric)}, \qquad (10.72)$$

$$\frac{1}{\sqrt{2}}[\psi_A(1)\psi_B(2) - \psi_A(2)\psi_B(1)][\beta(1)\beta(2)] \quad \text{(antisymmetric)}, \qquad (10.73)$$

$$\frac{1}{\sqrt{2}}[\psi_A(1)\psi_B(2) - \psi_A(2)\psi_B(1)]\frac{1}{\sqrt{2}}[\alpha(1)\beta(2) + \alpha(2)\beta(1)]$$
(antisymmetric), (10.74)

$$\frac{1}{\sqrt{2}}[\psi_A(1)\psi_B(2) - \psi_A(2)\psi_B(1)]\frac{1}{\sqrt{2}}[\alpha(1)\beta(2) - \alpha(2)\beta(1)]$$
(symmetric), (10.75)

THE PAULI EXCLUSION PRINCIPLE

Experimental evidence necessitates the postulation of an additional principle which further restricts the forms of allowed functions which describe the system. The postulate, first enunciated by Wolfgang Pauli in 1924, is called the *Pauli exclusion principle* and may be stated in its quantum- mechanical form as follows :

To be acceptable, the total wave function for a system of electrons must be antisymmetric to the simultaneous exchange of coordinates and spins between any pair of electrons.

According to the Pauli principle, only the *antisymmetric* complete functions given above are experimentally satisfactory for the two-electron system in which one electron is in the ψ_A orbital and one electron is in the ψ_B orbital. That is, the only satisfactory total functions are (10.72), (10.73), (10.74), and (10.75).

If one of the electrons is in the 1s hydrogen-like orbital and the other electron is in the 2s orbital, the four satisfactory total wave functions may be expressed as

$$\frac{1}{\sqrt{2}}[1s(1)2s(2) + 1s(2)2s(1)]\frac{1}{\sqrt{2}}[\alpha(1)\beta(2) - \alpha(2)\beta(1)], \qquad (10.76)$$

which contains the *symmetric spatial* function, and

$$\frac{1}{\sqrt{2}}\,[1s(1)2s(2) - 1a(1)2s(1)]\;[\alpha(1)\alpha(2)], \qquad (10.77)$$

$$\frac{1}{\sqrt{2}}\,[1s(1)2s(2) - 1a(1)2s(1)]\;[\beta(1)\beta(2)], \qquad (10.78)$$

$$\frac{1}{\sqrt{2}}\,[1s(1)2s(2) - 1s(2)2s(1)]\;\frac{1}{\sqrt{2}}\,[\alpha(1)\beta(2) + \alpha(2)\beta(1)], \qquad (10.79)$$

each of which contains the *antisymmetric spatial* function. The one allowed function (Eq. 10.76) which contains the symmetric spatial part is also associated with two electrons which have opposed spins. We say that the spins of such electrons are *paired.*

For the first of the total functions (Eq. 10.77) which contains the antisymmetric spatial part, note that the electron spins are coupled parallel and *up.* The net z-component of total spin angular momentum is

$$+\frac{\tfrac{1}{2}h}{2\pi} + \frac{\tfrac{1}{2}h}{2\pi} = \frac{(+1)h}{2\pi}.$$

For the second total function (Eq. 10.78) which contains the antisymmetric spatial part, we see that the electron spins are again coupled parallel but oriented down. The z-component of spin angular momentum is

$$-\frac{\tfrac{1}{2}h}{2\pi} - \frac{\tfrac{1}{2}h}{2\pi} = \frac{(-1)h}{2\pi}.$$

For the third total function (Eq. 10.79) which contains the antisymmetric spatial part, the net z-component of spin angular momentum is zero. This may be pictured as resulting when the z-component of two *coupled parallel* spins is zero.

We thus associate the symmetric space function with two electrons having opposed spins and the antisymmetric space functions with two electrons having parallel spins. When the electrons have opposed spins, they tend to be drawn together. When they have parallel spins, they tend to avoid each other. This interaction does not derive from a consideration of any direct force between the electrons. It arises entirely from our insistence that we account for the indistinguishability of electrons with respect to coordinates and intrinsic spins.

Let us now state the Pauli exclusion principle in a more familiar form. When two electrons in the same atom are in different spatial orbitals, that is, when any one of their space quantum numbers n, l, or m is different, they may have either the same spins or different spins (their m_s value may be the same or different). However, when both electrons are in the same spatial orbital, that is, when they have the same value for the quantum numbers, n, l, and m, the only way in which the complete wave function may be antisymmetric and thus conform to the Pauli principle is for the spin function to be antisymmetric, that is

$$(1/\sqrt{2})[\alpha(1)\beta(2) - \alpha(2)\beta(1)].$$

In such a case the complete wave function would be

$$\psi_A(1)\psi_A(2)(1/\sqrt{2})[\alpha(1)\beta(2) - \alpha(2)\beta(1)], \qquad (10.80)$$

and the electrons would thus be required to have opposed spins or different values for the spin quantum number m_s. The *Pauli exclusion principle* is usually sates as follows :

No two electrons in a given atom may have all four quantum numbers the same.

SINGLET AND TRIPLET STATES

The allowed complete wave functions for the two electrons in the 1s2s excited state of the helium atom are given by Eqs. (10.76) to (10.79). When the spatial function is symmetric, the only complete wave function which is satisfactory is the one (Eq. 10.76) in which the electron spins are opposed or paired. Since the symmetric spatial function appears in only one of the allowed states, it is said to be a *singlet* state. The singlet state is nondegenerate and has only one spin arrangement for the pair of electrons.

When the spatial function is antisymmetric, however, the complete wave functions given by Eqs. (10.77), (10.78), and (10.79) represent three degenerate states in which three different spin arrangements are possible. The threefold-degenerate state which involves the antisymmetric wave function is called a *triplet* state. The three states are degenerate in our present approximation because we are assuming no interaction between spin and orbital magnetic moments. Actually, there are small magnetic interactions between spin and orbital motions in the helium atom which split the triplet state into three closely spaced energy levels.

Whenever the spatial function for a pair of electrons in *any* two hydrogen-like orbitals is antisymmetric, any one of the three symmetric spin functions given by Eqs. (10.64), (10.65), and (10.66) may be chosen in conformity with the Pauli principle. Thus, antisymmetric two-electron spatial functions are always triplet states in which the electrons are unpaired. Symmetric two-electron spatiad functions, on the other hand, always represent singlet in which the electrons are paired. Since there are only two allowed spatial functions for the two-electron system, the symmetric function given by Eq. (10.62) and the antisymmetric function given by Eq. (10.63), we'll always expect a singlet state and a triplet state for each two-electron combination in which the hydrogen-like orbitals are different. If the two electrons are in the same orbital, the antisymmetric spatial function given by Eq. (10.63) vanishes to zero and the only allowed state is a nondegenerate singlet state. Some of the allowed states in the unperturbed helium atom are given in Table 10.1. These same allowed states apply to other *unperturbed* two-electron systems such as Li^+, Be^{2+}, B^{3+}, c^{4+}, etc.

TABLE 10.1

Some Allowed States in the Unperturbed Helium Atom

State		Symmetry of spatial function	Electron spins	Degeneracy
$1s^2$	singlet	Symmetric	Paired	1
$1s2s$	singlet	Symmetric	Paired	1
	triplet	Antisymmetric	Unpaired	3
$1s2p_x$	singlet	Symmetric	Paired	1
	triplet	Antisymmetric	Unpaired	3
$1s2p_y$	singlet	Symmetric	Paired	1
	triplet	Antisymmetric	Unpaired	3
$1s2p_z$	singlet	Symmetric	Paired	1
	triplet	Antisymmetric	Unpaired	3
$2s^2$	singlet	Symmetric	Paired	1

11

Many-Electron Atoms

In calculating the allowed energy levels for the hydrogen atom we noted that the energy of a given state is dependent entirely on the value of *n*, the principal quantum number. The lowest energy state, or ground state, is that state in which the lone electron is in the 1s orbital. The next highest energy level for the hydrogen atom may be represented by either the 2s or 2p state, since these two states are degenerate in hydrogen. Next in energy come the 3s, 3p, and 3d states, which are again degenerate in the hydrogen atom.

However the energies of different states of the two-electron *helium* atom namely the 1s2s and 1s2p states were not degenerate, primarily because of the difference in coulombic energies due to electron repulsion.

RELATIVE ENERGIES OF s, p, d AND r SUBLEVELS

A summary of some of the energies for the helium atom calculated through first-order perturbation theory is shown in Fig. 11.1. The energies shown are corrected for coulombic energies but not for exchange energies. This is equivalent to taking the simple (not weighted) mean of the energies of the singlet and triplet states for the 1s2s or 1s2p state. That is, the simple mean of the calculated energies for the singlet and triplet states is given as

$$\frac{(E^0 + J + K) + (E^0 + J - K)}{2} = E^0 + J.$$

As a result of the difference in the magnitude of electron repulsion energy, the 1s2s sate of the helium atom is a more stable state (a lower energy state) than the 1s2p state. If we were to extend a first-order perturbation treatment to the 1s3s, 1s3p, and 1s3d excited states of helium, we would find that the mean energies of the singlet and triplet states would follow the order

$$1s^2 < 1s2s < 1s2p < 1s3s < 1s3p < 1s3d.$$

increasing energy →

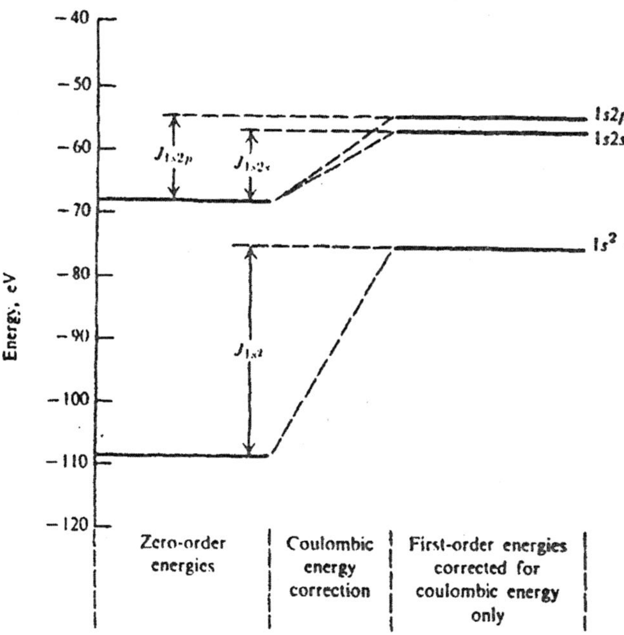

Fig. 11.1. Energies for three states of the helium atom, as calculated by first-order perturbation theory. No corrections are included for exchange energies for the 1s2s and 1s2p states; that is, the energies of the singlet and triplet states are averaged.

That is, if the first electron is in the 1s state, the states of increasing energy for the second electron in the helium atom are

$$1s < 2s < 2p < 3s < 3p < 3d.$$

In more complex atoms containing larger numbers of electrons, the coulombic and exchange interactions become more complex and the relative positions of the energies of the various one-electron orbitals will change. Nevertheless, it will be seen that within a given major energy level, that is, for a given value of the principal quantum number n, the order of increasing energy within sublevels is always given as

$$s < p < d < f.$$

SELF-CONSISTENT FIELD (SCF) METHOD FOR MANY-ELECTRON ATOMS

Although it is easy to *write* the Schrodinger equation for any given many electron atom, a direct solution is impossible, even in the relatively simple case of the helium atom. The specific terms which make separation of variables impossible are the electronic repulsion terms in the potential-energy portion of the Hamiltonian operator.

Because of the lack of a direct formal solution in finite terms, it is necessary to resort to approximation methods for complex atoms. But even if we could solve the Schrodinger equation exactly for polyelectronic atoms, there are severe practical limitations.

Hartree in 1928 proposed a successive approximation method known as the *self-consistent field (SCF) method for solving* the Schrodinger equation for a polyelectronic atom. In this method we first ignore interelectron repulsion and define a tentative normalized wave function φ for each electron. We then assume that the total nonperturbed wave function for an atom containing n electrons may be expressed as

$$\psi = \varphi_1(1)\varphi_2(2)\varphi_3(3) \ldots\ldots \varphi_n(n), \qquad (11.1)$$

where electron 1 is in the φ_1 orbital, electron 2 is in the φ_2 orbital, etc. This, of course, is the same type of function we used in approximating the ground-state energy of the unperturbed helium atom.

In order to select reasonably accurate *tentative* one- electron orbitals φ_1 to use in Eq. (11.1), we initially assume that each of the electrons moves in the *same* central *spherically symmetric* potential-energy field $V(r)$, so that we may write the same one- electron Schrodinger equation for each electron. That is, assuming a stationary nucleus, we write

$$\frac{-h^2}{8\varphi^2 m_e}\nabla^2\varphi + V(r)\varphi = E\varphi, \qquad (11.2)$$

in which $V(r)$ is a first rough approximation of the net potential energy field in the polyelectronic atom. In the Schrodinger equation for the single electron in the hydrogen-like ion, $V(r)$ was given as $-Ze^2/r$. However, in constructing $V(r)$ for use in Eq. (11.2), we attempt to take into account the spherically averaged effect of interaction with other electrons, using *numerical* methods, so that the resultant $V(r)$ is usually presented as a table or as a plot. Thus, Eq. (11.2) usually must be integrated by numerical methods. After obtaining the set of allowed φ functions and the corresponding eigenvalue energies by numerical solution of Eq. (11.2), we assign tentative orbitals to each of the n individual electrons in order of increasing energy, always conforming to the Pauli exclusion principle which limits the number of electrons in each spatial orbital to two (one with spin up and one with spin down). At this point in the calculation, the forms of the individual one-electron functions φ_i and the corresponding energies E_i are only rough approximations, due primarily to the inaccuracy of the initial potential energy term $V(r)$ which was used in the Eq. (11.2).

In order to imporve the accuracy of the potential-energy term we now assume that for any *one selected electron*, the entire potential energy of interaction with the nucleus and with all of the remaining electrons may be expressed in terms of a *spherically symmetric* potential-energy function, V_i *different for each electron,* which at any point in space represents the sum of the potential energy of attraction to the nucleus and the potential energies of repulsion between the selected electron and each of the remaining electrons. The spherical symmetry of the potential energy term V_i is important because the Schrodinger equation may be solved directly *only* when V_i is a function of r_i alone, where r_i is the distance between the nucleus and the electron under consideration. In addition the assumption of spherical symmetry is reasonably accurate because the potential energy of interaction between a given electron and the nucleus is spherically symmetric and because the total charge distribution for any *filled* l level (closed s, p, d, or f shell) may be shown to be spherically symmetric; that is, charge distributions in configurations such as 1s or $1s^2 2p^6$ are spherically symmetric. In the Hartree method we further average the remaining nonspherical charge distributions over all angles to ensure spherical symmetry for the entire potential-energy function and assume that the resultant error will be small.

We can now write the independent Schrodinger equation for electron 1 as

$$\frac{-h^2}{8\pi^2 m_e}\nabla_1^2\varphi_1 + V_1\varphi_1 = E_1\varphi_1, \tag{11.3}$$

where V_1 depends only on r_1. For the ith electron we can write

$$\frac{-h^2}{8\pi^2 m_e}\nabla_i^2\varphi_i + V_i\varphi_i = E_i\varphi_i, \tag{11.4}$$

where V_i depends only on r_i. The total energy of the atom is given as

$$E = \sum_{i=1}^{n} E_i \tag{11.5}$$

However, in order to solve Eq. (11.4), we must determine how we'll formulate the spherically symmetric potential-energy term V_i. For this, we first consider the interaction between a selected electron, electron 1 for example, and one of the other remaining electrons, say electron 2. Let us imagine that electron 2 is smeared out into a charge distribution such that at any point in space the fraction of the electron which is contained in a unit volume is given by $|\varphi_2|^2$ and the charge per unit volume is $e|\varphi_2|^2$. Then the charge in the differential volume element $d\tau_2$ is given as follows

$$e\,|\varphi_2|^2\,d\tau_2 = e\,|\varphi_2|^2 dx_2\,dy_2\,dz_2. \tag{11.6}$$

The total potential energy of repulsion between electron 1 at a given point (x_1, y_1, z_1) and electron 2 is obtained by integrating the interaction of the charge e of electron 1 with each differential charge of electron 2 (Eq. 11.6) over all space, and is written as follows

$$\iiint \frac{e\cdot e\,|\varphi_2|^2\,dx_2\,dy_2\,dz_2}{r_{12}} = e^2\iiint \frac{|\varphi_2|^2}{r_{12}}\,dx_2\,dy_2\,dz_2. \tag{11.7}$$

The complete potential energy term for electron 1 due to attraction to the nucleus of charge Z and to repulsion of *all of the other electrons* is then

$$V_1(x_1, y_1, z_1) = \frac{-Ze^2}{r_1} + e^2\sum_{i\neq 1}\iiint \frac{|\varphi_i|^2}{r_{1i}}\,dx_i\,dy_i\,dz_i. \tag{11.8}$$

The solution of Eq. (11.8) for all closed shell systems leads to a spherically symmetric function for V_1. For atoms containing electrons in open shells, the summation term of Eq. (11.8) must be averaged over all angles, so that V_1 will be spherically symmetric. In either case, once a spherically symmetric form for V_1 is obtained, it is then used in solving Eq. (11.3) to evaluate a first value for E_1 and a first-improved version of φ_1.

An identical calculation is then repeated for each of the other electrons. That is, for each of the other electrons a potential energy field V_i is calculated from the expression

$$V_i(x_i, y_i, z_i) = \frac{-Ze^2}{r_i} + e^2 \sum_{j \neq i} \iiint \frac{|\varphi_j|^2}{r_{ij}} \, dx_j \, dy_j \, dz_j. \qquad (11.9)$$

which is then used in Eq. (11.4) in order to evaluate a first energy E_i and a first-improved version of φ_i. At this point, we would have a complete set of first-improved wave functions, one for each electron.

The next step is to repeat each of the above calculations, using the set of first-improved φ functions, in order to calculate a new set of energies and a new set of second-improved wave functions. We then again repeat the process a third time, this time using the second-improved φ functions in order to calculate third-improved energies and third-improved φ functions. This process is repeated until there are no further appreciable changes in the E_i values, at which point the φ functions are said to be *self-consistent*, and the energy of the atom is then given by Eq. (11.5).

Since the Schrodinger equation, Eq. (11.4), for each of the one-electron functions is identical in form to the Schrodinger equation for the single electron in the hydrogen-like ion, the mechanism of solution of Eq. (11.4) is the same as that for the hydrogen-like ion. The two Schrodinger equations differ only with respect to the forms of the spherically symmetric potential- energy term V_i. Moreover, since V_i appears *only* in the equation for the radial solution, *the solutions for* $\Phi(\varphi_i)$ and $\Theta(\theta_i)$ *are exactly the same as for the hydrogen-like ion,* which means that the angular shapes and probability distributions for the one-electron φ functions are identical to those for the single electron in the hydrogen-like ion. That is, the individual electrons in a poly-electronic atom may be represented by the same angular distributions for s, p, d, and f orbitals as done earlier.

On the other hand, the radial functions for the individual electrons $R'(r_i)$ are *not* the same as those for the single electron in a hydrogen-like ion, nor are the corresponding eigenvalue energies E_i, which are obtained from the solution of the one-electron radial equations, the same.

For each of the one-electron functions, we can write

$$\varphi_i = R'_{nl}(r_i)\Theta_{lm}(\theta_i)\Phi_m(\varphi_i),$$

where Θ_{lm} and Φ_m are identical to the corresponding functions for the hydrogen-like ion, respectively, and where R'_{nl} is different for each electron and is obtained numerically as a table or plot, rather than as an explicit equation. The final wave equation ψ for the entire atom of n electrons is then given according to Eq. (11.1) as the product of n individual φ functions as

$$\psi = \prod_{i=1}^{i-n} \varphi_i = \prod_{i=1}^{i=n} [R'_{nl}(r_i)\Theta_{lm}(\theta_i)\Phi_m(\varphi_i)].$$

The Hartree self-consistent field method, which is simple in concept but bulky and cumbersome in computation, has four limitations.

1. The expression of ψ as a product of φ functions for each of the electrons ignores the problem of electron indistinguishability. The integrals in Eqs. (11.7), (11.8), and (11.9) are *coulombic* integrals. To be more accurate we should write ψ in a form which ensures antisymmetry. Such an antisymmetric wave function introduces *exchange* integrals and energies. Calculations based on antisymmetric determinant forms of ψ yield somewhat improved energy levels and wave functions.

2. The method does not consider the energies involved in spin-orbital interactions, which are important when one deals with atomic spectra.

3. Many atoms contain electrons which are not part of closed l subshells, so that the potential-energy field is not truly spherically symmetrical and a resultant error is introduced.

4. No provision is made to include the energies associated with the instantaneous correlation of relative elctron positions.

GROUND-STATE ENERGY LEVELS IN COMPLEX ATOMS

The SCF method may be used to calculate the energy for each of the electrons in a polyelectronic atom. For example, the ground state configuration of the carbon atom is $1s^2 2s^2 2p^2$ and the total energy of the atom in the ground state is given by

$$E = 2E_{1s} + 2E_{2s} + 2E_{2p}, \tag{11.10}$$

where E_{1s}, E_{2s}, and E_{2p} are the energies of the electrons in the *carbon atom* orbitals indicated by the subscripts. Then, we might calculate the individual orbital energies for each of the elements in order to obtain an overall picture of atomic behavior. However, SCF calculations are complicated by the lengthy numerical integrations which are required to obtain the self-consistent solutions.

Energy calculations may be very much simplified if we assume that each of the electrons moves independently in the *same* central potential-energy field. The resulting orbital energies, agree in general with experimental energies as well as with energies calculated by the more complicated SCF method. As the nuclear charge Z increases, the energies of the principal levels begin to split into separate energies in each of the l sublevels. The observed energy splitting is due to nuclear shielding effects and to differences in relative penetrations among the orbitals in multi-electronic systems. For example, the 3s orbital is lower in energy than the 3p orbital because the 3s electron is more easily able to penetrate the screen of 1s and 2s electrons than is the 3p electron. In general the lower the l quantum number for an orbital, the more penetrating is the orbital. On the other hand, orbitals, having a lower value of l are less effective in screening than orbitals of higher l quantum number. At intermediate Z values, the order of relative orbital energies becomes fairly complicated, and crossovers in energy levels occur which account for many of the periodic properties of the elements. At high values for Z, the l sublevels within a given n principal level again coverage, apparently due to the overshadowing effect of the very high nuclear charge which tends to reduce differences in energies between l suborbitals.

ELECTRONIC DISTRIBUTIONS IN COMPLEX ATOMS

We will now consider *an imaginary process by which we construct electronic configurations for each of the elements by adding electrons to the atom one at a time* (of course protons and neutrons are added simultaneously). In constructing an atom according to such an *Aufbau* (German for *build-up*) procedure we follow three guidelines:

1. *Electrons must always be placed in orbitals of lowest possible energy*
2. *the Pauli principle, requires that no two electrons may have the same set of four quantum numbers, and*
3. *Hund's rule of maximum multiplicity requires that, other things being equal (for example, in partially filled degenerate orbitals), the state with the maximum number of unpaired electrons be lowest in energy.*

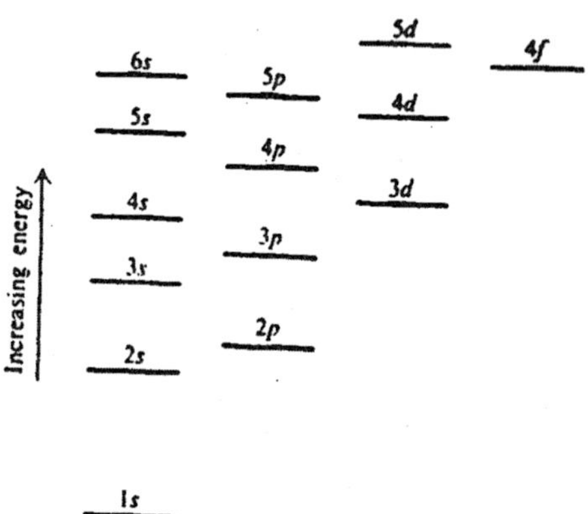

Fig. 11.2. Relative orbital energies *in the order in which atomic orbitals are filled* by successive addition of single electrons according to the Aufbau process.

Although the relative energies of orbitals vary with Z, we may construct a chart of the energies of orbitals *in order in which they would be filled* by successive addition of electrons, beginning with the hydrogen atom and continuing through the elements to the most complex atom (Fig. 11.2). Each of the s orbitals may contain one pair of electrons, opposed in spin. For the p level, however, $l = 1$ and m may have the values -1, 0, $+1$, so that each p level contains three degenerate *m* orbitals, each of which may contain one pair of electrons. Similarly, for each d level, $l = 2$, and *m* may have the values −2, −1, 0, +1, +2. Each of the five degenerate *m* orbitals may in turn contain a pair of electrons, one with spin up and one with spin down.

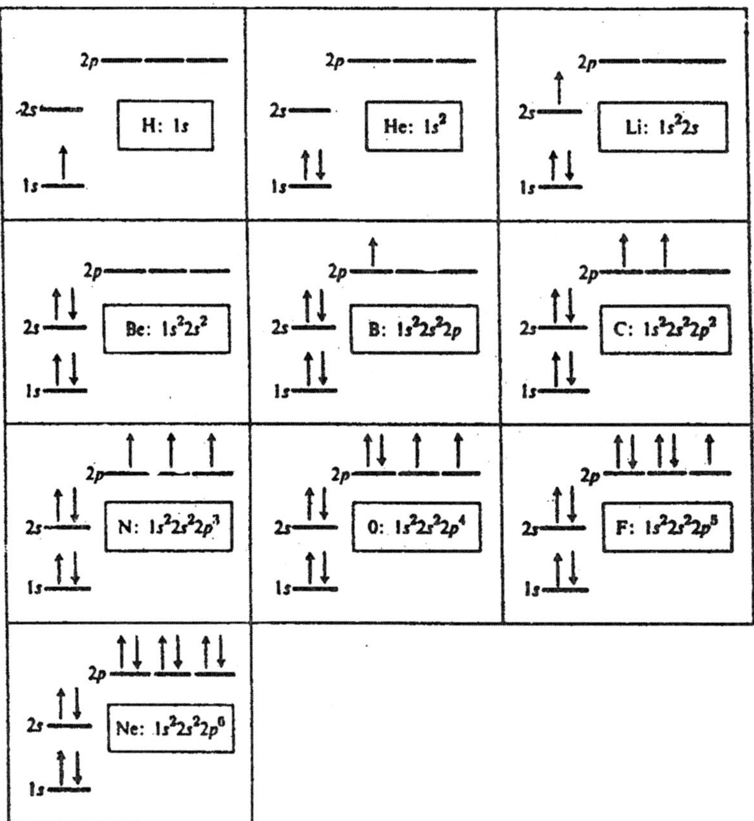

Fig. 11.3. Electronic configuration for the first 10 elements.

Thus each d level may contain a total of 10 electrons or five degenerate pairs. Similarly, each f level may contain a total of 14 electrons, or seven degenerate pairs.

We can now use the energy chart given in Fig. 11.2 in conjunction with the Pauli principal and Hund's rule in order to construct the electronic configurations of the elements. The relative electron spin is usually indicated by arrows pointed up or down. The configuration for the first 10 elements are shown in Fig. 11.3. Note that when the configuration for carbon in formed by adding an electron the configuration for boron, the two electrons in the 2p orbitals have parallel spins according to Hund's rule. for nitrogen, the three 2p electrons also have parallel spins. As a general rule, when p levels are filled, the first three electrons are placed with parallel spins in separate orbitals. Similarly, when d levels are filled, the first five electrons have parallel spins in separate orbitals, and when f levels are filled, the first seven electrons are placed with parallel spins in separate f orbitals.

Electronic configurations of all of the elements are presented in Table 11.1. In interpreting these configurations we should always keep Hund's rule in mind. For example, the configuration for Ru (element 44) is given as [Kr]$4d^7 5s$, so that the configuration of the 4d-level electrons may be interpreted as

$$4d \ \underline{\uparrow\downarrow} \ \underline{\uparrow\downarrow} \ \underline{\uparrow} \ \underline{\uparrow} \ \underline{\uparrow}$$

That is, three of the 4d electrons are unpaired and have parallel spins. Note also that the 5s electron in Ru is unpaired. All of the remaining electrons in the [Kr] kernel are paired.

Although the observed configurations agree fairly well with predictions, there are a few notable exceptions. Consider configurations (tan;e 11.2) for the group of elements from Sc through Zn. These are the elements whose configurations are constructed as we fill in the 10 electrons in the 3d orbital. This group of elements is known as the *first transition series.* Note that the configurations of both Cr and Cu are apparent exceptions to the general rule. For each of these atoms, one 4s electron has unpaired and entered the higher 3d energy level in order to either exactly half-fill or completely fill the 3d energy level, with an apparent increase in stabilization (lowering of energy) of the atom. In each case, the net lowering of energy may be explained in terms of the exchange energy, K, which we have earlier defined for a set of *two* electrons of parallel spin. The exchange energy is the lowering of energy for a set of two electrons of parallel spin which arises as a result of the requirement of indistinguishability or exchangeability of the electron coordinates. Two electrons with parallel spins remain farther apart than two electrons with opposed spins and consequently the coulombic repulsion energy is lower. In a system containing more than two electrons with parallel spins, the lowering of

TABLE 11.1
Electronic Configurations of Atoms in the Gaseous State

Atomic number	Element	Electronic configuration	Atomic number	Element	Electronic configuration
1	H	1s	53	I	$-4d^{10}5s^2 5p^5$
2	He	$1s^2$	54	Xe	$-4d^{10}5s^2 5p^6$
3	Li	[He] $2s^1$	55	Cs	[Xe] $6s^1$
4	Be	$-2s^2$	56	Ba	$-6s^2$
5	B	$-2s^2 2p^1$	57	La	$-5d^1 6s^2$
6	C	$-2s^2 2p^2$	58	Ce	$-4f^2 6s^2$
7	N	$-2s^2 2p^3$	59	Pt	$-4f^3 6s^2$
8	O	$-2s^2 2p^4$	60	Nd	$-4f^4 6s^2$
9	F	$-2s^2 2p^5$	61	Pm	$-4f^5 6s^2$
10	Ne	$-2s^2 2p^6$	62	Sm	$-4f^6 6s^2$
11	Na	[Ne] $3s^1$	63	Eu	$-4f^7 6s^2$
12	Mg	$-3s^2$	64	Gd	$-4f^7 5d^1 6s^2$
13	Al	$-3s^2 3p^1$	65	Tb	$-4f^9 6s^2$
14	Si	$-3s^2 3p^2$	66	Dy	$-4f^{10} 6s^2$
15	P	$-3s^2 3p^3$	67	Ho	$-4f^{11} 6s^2$
16	S	$-3s^2 3p^4$	68	Er	$-4f^{12} 6s^2$
17	Cl	$-3s^2 3p^5$	69	Tm	$-4f^{13} 6s^2$
18	Ar	$-3s^2 3p^6$	70	Yb	$-4f^{14} 6s^2$
19	K	[Ar] $4s^1$	71	Lu	$-4f^{14} 5d^1 6s^2$
20	Ca	$-4s^2$	72	Hf	$-4f^{14} 5d^2 6s^2$
21	Sc	$-3d^1 4s^2$	73	Ta	$-4f^{14} 5d^3 6s^2$
22	Ti	$-3d^2 4s^2$	74	W	$-4f^{14} 5d^4 6s^2$
23	V	$-3d^3 4s^2$	75	Re	$-4f^{14} 5d^5 6s^2$
24	Cr	$-3d^5 4s^1$	76	Os	$-4f^{14} 5d^6 6s^2$
25	Mn	$-3d^5 4s^2$	77	Ir	$-4f^{14} 5d^7 6s^2$
26	Fe	$-3d^6 4s^2$	78	Pt	$-4f^{14} 5d^9 6s^1$
27	Co	$-3d^7 4s^2$	79	Au	$4f^{14} 5d^{10}$
28	Ni	$-3d^8 4s^2$	80	Hg	$-6s^2$
29	Cu	$-3d^{10} 4s^1$	81	Tl	$-6s^2 6p^1$
30	Zn	$-3d^{10} 4s^2$	82		$-6s^2 6p^2$

TABLE 11.1
continued....

Atomic number	Element	Electronic configuration	Atomic number	Element	Electronic configuration
31	Ga	$-3d^{10}4s^24p^1$	83	Bi	$-6s^26p^3$
32	Ge	$-3d^{10}4s^24p^2$	84	Po	$-6s^26p^4$
33	As	$-3d^{10}4s^24p^3$	85	At	$-6s^26p^5$
34	Se	$-3d^{10}4s^24p^4$	86	Rn	$-6s^26p^6$
35	Br	$-3d^{10}4s^24p^5$	87	Fr	[Rn] 7s
36	Kr	$-3d^{10}4s^24p^6$	88	Ra	$-7s^2$
37	Rb	[Kr] $5s^1$	89	Ac	$-6d^17s^2$
38	Sr	$-5s^2$	90	Th	$-6d^27s^2$
39	Y	$-4d^15s^2$	91	Pa	$-5f^26d^17s^2$
40	Zr	$-4d^25s^2$	92	U	$-5f^36d^17s^2$
41	Nb	$-4d^45s^1$	93	Np	$-5f^46d^17s^2$
42	Mo	$-4d^55s^1$	94	Pu	$-3f^67s^2$
43	Tc	$-4d^55s^2$	95	Am	$-5f^77s^2$
44	Ru	$-4d^75s^1$	96	Cm	$-5f^76d^17s^2$
45	Rh	$-4d^85s^1$	97	Bk	$-5f^97s^2$
46	Pd	$-4d^{10}$	98	Cf	$-5f^{10}7s^2$
47	Ag	$-4d^{10}5s^1$	99	Es	$-5f^{11}7s^2$
48	Cd	$-4d^{10}5s^2$	100	Fm	$-5f^{12}7s^2$
49	In	$-4d^{10}5s^25p^1$	101	Md	$-5f^{13}7s^2$
50	Sn	$-4d^{10}5s^25p^2$	102	No	$-5f^{14}7s^2$
51	Sb	$-4d^{10}5s^25p^3$	103	Lw	$5f^{14}6d^17s^2$
52	Te	$-4d^{10}5s^25p^4$			

energy in the ground state due to exchange (the total exchange energy stabilization) is given by the product of the total number of possible exchange of coordinates between sets of two electrons of parallel spin times the exchange energy K per set of electrons of parallel spin. For example, for system containing three electrons, 1, 2, and 3, all having parallel spins, there are three possible exchanges of coordinates as follows :

1 exchanges with 2,

1 exchanges with 3,

2 exchanges with 3.

TABLE 11.2
Configurations for the Elements in the First Transition Series

Element	Electron configuration	
	Predicted	Observed
Sc	$[Ar]3d^14s^2$	$[Ar]3d^14s^2$
Ti	$-3d^24s^2$	$-3d^24s^2$
V	$-3d^34s^2$	$-3d^34s^2$
Cr	$(-3d^44s^2)$*	$-3d^54s^1$
Mn	$-3d^54s^2$	$-3d^54s^2$
Fe	$-3d^64s^2$	$-3d^64s^2$
Co	$-3d^74s^2$	$-3d^74s^2$
Ni	$-3d^84s^2$	$-3d^84s^2$
Cu	$(-3d^94s^2)$*	$-3d^{10}4s^1$
Zn	$-3d^{10}4s^2$	$-3d^{10}4s^2$

*Exceptions.

Thus the total exchange energy stabilization is $3K$. For a system containing four electrons, 1, 2, 3, and 4, all having parallel spins, there are six possible exchanges of coordinates positions among electrons as follows :

1 exchanges with 2,

1 exchanges with 3,

1 exchanges with 4,

2 exchanges with 3,

2 exchanges with 4,

3 exchanges with 4,

and the total exchange energy stabilization is $6K$. In general, it may be shown that for a system containing n electrons of parallel spin, the number of possible sets, each containing two electrons of parallel spin, is given as

$$n!/2(n-2)!.$$

Recognizing that each set of two electrons of parallel spin contributes an energy K to the total exchange energy stabilization, we may write the total exchange energy stabilization for each of the d level configurations as shown in Table 11.3.

TABLE 11.3
Total Exchange Energy Stabilization for d Level Configurations

Configuration	Number of possible sets of electrons of parallel spins	Total exchange-energy stabilization
d^1: ↑ _ _ _ _	0	0
d^2: ↑ ↑ _ _ _	1	K
d^3: ↑ ↑ ↑ _ _	3	$3K$
d^4: ↑ ↑ ↑ ↑ _	6	$6K$
d^5: ↑ ↑ ↑ ↑ ↑	10	$10K$
d^6: ↑↓ ↑ ↑ ↑ ↑	10	$10K$
d^7: ↑↓ ↑↓ ↑ ↑ ↑	11 (10 up, 1 down)	$11K$
d^8: ↑↓ ↑↓ ↑↓ ↑ ↑	13 (10 up, 3 down)	$13K$
d^9: ↑↓ ↑↓ ↑↓ ↑↓ ↑	16 (10 up, 6 down)	$16K$
d^{10}: ↑↓ ↑↓ ↑↓ ↑↓ ↑↓	20 (10 up, 10 down)	$20K$

In order to explain the unexptected configuration for the chromium atom, we note that in the absence of exchange effects, the $3d^44s^2$ state would be slightly lower in energy (slightly more stable) than the $3d^54s$ state. However, because of the greater possibility for electron exchange in the $3d^54s^1$ state, its energy is lowered by $10K$, whereas the lowering of energy due to electron exchange in the $3d^44s^2$ state is only $6K$. The net result is that the observed energy level of the $3d^54^s$ state is *lower* than that of the $3d^44s^2$ state. The unexpected stability of the $3d^{10}4s$ configuration for the copper atom may be explained in the same way. As would be expected, it can be observed that exactly half-filled or completely filled p shells and f shells also show unusual stability, because of exchange effects.

SECTION—IV

PROBLEMS AND SOLUTIONS

Problems and Solutions

PROBLEM–1

How will you solve a simple differential equation of the second order of the following form

$$\frac{d^2y}{dx^2} + k^2y = 0 \tag{A1.1}$$

SOLUTION

In Eq. (A1.1) k^2 is constant. Such an equation is satisfied by a function of an exponential form. Let us assume the solution of Eq. (A1.1) as

$$y = \exp(mx) \tag{A1.2}$$

Then substituting for y, $\exp(mx)$ in Eq. (A1.1), we obtain the auxiliary equation

$$m^2 + k^2 = 0 \tag{A1.3}$$

The roots of Eq. (A1.3) are given by

$$m = \pm ik \tag{A1.4}$$

where $i = \sqrt{-1}$. The particular solutions of Eq. (A1.1) are $\exp(+ikx)$ and $\exp(-ikx)$. On adding these particular solutions there results a solution with two independent arbitrary constants and therefore the complete solution of Eq. (A1.1) is written as

$$y = A_1 \exp(ikx) + A_2 \exp(-ikx) \tag{A1.5}$$

On expanding the exponentials in sines and cosines we obtain the following equivalent form

$$y = A_1 (\cos kx + i \sin kx) + A_2 (\cos kx - i \sin kx)$$

$$= (A_1 + A_2) \cos kx + (A_1 - A_2)\, i \sin kx$$

$$= C \cos kx + D \sin kx \qquad \text{(A1.6)}$$

In Eqs. (A1.5) and (A1.6), A_1, A_2, C, and D are arbitrary constants.

PROBLEM–2

Using the Heisenberg's uncertainty principle calculate the ground-state energy, E_0, of the helium atom of nuclear charge $+Ze$.

SOLUTION

The helium atom contains two electrons. The total energy, E, is the sum of the kinetic energy, T; the potential energy of attraction of the two electrons towards the nucleus, $V(r_1, r_2)$; and the interelectronic repulsion energy, $V(r_{12})$:

$$E = T + V(r_1, r_2) + V(r_{12}) \qquad (1)$$

where

$$T = \frac{p^2}{2m} = \frac{p_1^2 + p_2^2}{2m}$$

$$V(r_1, r_2) = -Ze^2 \left(\frac{1}{r_1} + \frac{1}{r_2} \right)$$

$$V(r_{12}) \approx \frac{e^2}{r_{12}} \approx \frac{e^2}{r_1 + r_2}$$

Here r_1 is the distance of the first electron from the nucleus; r_2 the distance of the second electron from the nucleus; and r_{12}, the interelectronic distance. Proceeding as in the last problem,

$$p_1 \approx \Delta p_1 \approx \frac{\hbar}{r_1}$$

$$p_2 \approx \Delta p_2 \approx \frac{h}{r_2}$$

$$T \approx \frac{\hbar^2}{2m} \left(\frac{1}{r_1^2} + \frac{1}{r_2^2} \right)$$

$$E = \frac{h^2}{2m}\left(\frac{1}{r_1^2} + \frac{1}{r_2^2}\right) - Ze^2\left(\frac{1}{r_1} + \frac{1}{r_2}\right) + \frac{e^2}{r_1 + r_2} \tag{2}$$

In the ground-state the energy has its minimum value, E_0, so that:

$$\frac{\partial E}{\partial r_1} = -\frac{h^2}{mr_1^3} + \frac{Ze^2}{r_1^2} - \frac{e^2}{(r_1 + r_2)^2} = 0 \tag{3a}$$

$$\left(\frac{\partial E}{\partial r_2}\right) = -\frac{h^2}{mr_2^3} + \frac{Ze^2}{r_2^2} - \frac{e^2}{(r_1 + r_2)^2} = 0 \tag{3b}$$

Eqs. (3a) and (3b) are two simultaneous equations in the two unknowns r_1 and r_2 which, on solution, give

$$r_1 = r_2 = \frac{h^2}{me^2}\left(\frac{1}{Z - ¼}\right) \tag{4}$$

Substituting for r_1 and r_2 in Eq. (2),

$$E_0 \approx -\left(Z - \frac{1}{4}\right)^2 \frac{me^4}{h^2} = -\left(Z - \frac{1}{4}\right)^2 \times 27.2\ eV.$$

$$= 83.3 \text{ eV} \qquad (\because Z = 2)$$

The axperimental value of E_0 is –79.00 *eV*.

PROBLEM–3

Using uncertainty principle, estimate the kinetic energy of a nucleon (proton or neutron) in a carbon nucleus of radius 1.5 fermi (i.e., 1.5×10^{-15} m).

SOLUTION

Since the nucleon can be found anywhere in the nucleus, the uncertainty in its position is equal to the diameter of the nucleus, i.e.,

$$\Delta x \approx 3 \times 10^{-15} \text{ m}$$

Hence, from the uncertainty principle,

$$\angle px \approx \frac{h}{\Delta x} \approx 3 \times 10^{-20}\ kgm\ s^{-1} = \Delta py = \Delta pz.$$

$$p \approx \Delta p = \Delta px + \Delta py + \Delta pz = 9 \times 10^{-20}\ kgm\ s^{-1}.$$

Hence, the kinetic energy is :

$$T = \frac{p^2}{2m} \approx \frac{(\Delta p)^2}{2m} \approx \frac{3h^2}{2m(\Delta x)^2} \approx 7\ MeV.$$

PROBLEM–4

Suppose we consider a neutron as composed of a proton and an electron separated by a distance of 1 fermi (i.e., 1×10^{-15} m), held together by Coulombic interaction. Ascertain whether or not this model of the neutron is acceptable on the basis of the Uncertainty principle.

SOLUTION

See Fig. 1. $\Delta x \,.\, \Delta p_x \approx h$

$$\Delta x = 1 \text{ fermi} = 1 \times 10^{-15}\ m$$

$$\Delta p_x \approx \frac{h}{\Delta x} \approx \frac{1.05 \times 10^{-34}\ J\ s}{1 \times 10^{-15} m}$$

$$\approx 10^{-19} \text{ kg}\ m\ s^{-1}$$

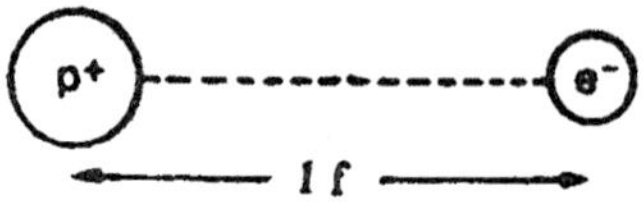

p+ ---------------- e-

← 1 f →

Fig. 1.

We have similar expression for the uncertainties in the y- and z-components, Δp_y and Δp_z, respectively. Hence the kinetic energy of the neutron is :–

$$T = \frac{p^2}{2M_n} \approx \frac{(\Delta p)^2}{2M_n} \approx \frac{(\Delta p_x)^2 + (\Delta p_y)^2 + (\Delta p_z)^2}{2M_n}$$

where M_n is the mass of the neutron.

T is calculated to be $\approx 2 \times 10^{-8}\ J$.

The Coulombic potential energy between the proton and the electron is :

$$V(r) = \frac{e^2}{r} = \frac{e^2}{(4\pi\varepsilon_0)r} \text{ (in } S.I. \text{ units)}$$

$$= \frac{(1.602 \times 10^{-19}\ C)^2}{(1.1126 \times 10^{-10}\ C^2 N^{-1} m^{-2})(1 \times 10^{-15} m)}$$

$$\approx 1.5 \times 10^{-13} Nm = 1.5 \times 10^{-13} J$$

Thus, we see that T » V(r). Hence, the given model of the neutron is not acceptable. The proton-electron system will fly apart, the potential energy being not sufficient to counterbalance the enormous kinetic energy.

PROBLEM–5

Using Wilson-Sommerfeld quantization rule determine :

(a) The energy levels of a particle of mass m which is constrained to move in a one-dimensional box where the potential energy V (x) is infinite at the walls and is zero inside it, i.e., V (x) = 0; 0<x<a, where a is the length of the box;

(b) the energy levels of a rigid rotor constrained to move about a fixed axis in a plane;

(c) the energy levels of a one-dimensional simple harmonic oscillator (S.H.O.).

SOLUTION

This problem deals with results derived from the so-called "Old Quantum Theory" which fell into disfavour after the discovery of Quantum Mechanics in 1926. According to the Wilson- Sommerfeld Quantization Rule,

$$J_i = \oint p_i \, dq_i = n_i h; \ (n_i = 0, 1, 2, 3, ...)$$

where the phase integral J_i extends over one period of the variable q_i. Here q and p are the generalized coordinate and the canonically conjugate momentum, respectively, of the moving particle.

(a) For a particle-in-a box (Fig. 2),

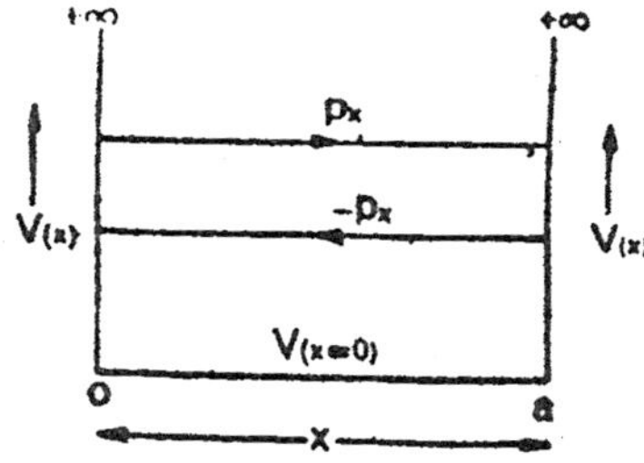

Fig. 2.

$$p_i \equiv p_x; q_i \equiv x.$$

As the particle moves, its momentum, p_x remains constant and equals $-p_x$ after it is reflected from the wall. Hence, the phase integral becomes :

$$J = \oint p_x\, d_x = \int_0^{2a} 2p_x dx = 2p_x \int_0^{a} dx' = 2p_x a = n_x h$$

$$p_x = \frac{n_x h}{2a}$$

The total energy E, which in this case equals the kinetic energy, T (because the potential energy inside the box is zero) is :

$$E \equiv En_x = \frac{p_x^2}{2m} = \frac{nx^2 h^2}{8ma^2}; (n_x = 1, 2, 3, ...)$$

(b) when the rigid rotator (rotor) moves about an axis in a plane its angular momentum, p_ϕ is given by

$$p_\phi = I_\phi$$

where ϕ is the angle of rotation; and I, the moment of inertia. Potential energy, $V = 0$. Hence the total energy E = the kinetic energy. Since angular momentum is a constant of the motion, we have

$$J = \oint p_\phi d\phi = p_\phi \int_0^{2\pi} d_\phi = 2\pi p_\phi = nh$$

$$p_\phi = n\left(\frac{h}{2\pi}\right) = nh$$

$$E \equiv En = .\frac{p^2\phi}{2I} = \frac{n^2 h^2}{2I}$$

It is customary to use the symbol J for the rotational number ($J \equiv n$) :

$$E_j = \frac{j^2 h^2}{2I}.$$

(c) The one dimensional S.H.O. moves under the Hooke's law potential V $(x) = ½\, kx^2$, where k is the force constant and x, the displacement from equilibrium configuration. The vibrational frequency of the oscillator is given by :

$$\nu = \frac{1}{2\pi}\left(\frac{k}{\mu}\right)^{1/2}$$

where μ is the reduced mass of the S.H.O. Its angular frequency, ω is :

$$\omega = 2\pi\nu = \left(\frac{k}{\mu}\right)^{1/2}$$

$$k = \mu\omega^2$$

$$V(x) = \tfrac{1}{2}\,\mu\omega^2 x^2$$

The kinetic energy, $T = p^2_x/2m$. Hence the total energy is

$$E = \frac{p_x^2}{2\mu} + \tfrac{1}{2}\,\mu\omega^2 x^2$$

$$= \tfrac{1}{2}\,\mu\omega^2 A^2 \qquad (1)$$

where A is the amplitude of the S.H.O.

From Eq. (1),

$$p_x = \mu\omega\,(A^2 - x^2)^{1/2} \qquad (2)$$

Also, we can write

$$x = A\cos\phi$$

$$dx = -A\sin\phi\, d\phi$$

$$(A^2 - x^2)^{1/2} = \sqrt{A^2(1-\cos^2\phi)} = A\sin\phi$$

$$p_x = A\,\mu\omega\sin\phi$$

Hence, the phase integral is :

$$J = \oint p_x\,dx \rightarrow \int 2p_x\,dx \rightarrow 2A^2\,\mu\omega\int_0^{\pi}\sin^2\phi\,d\phi$$

$$= A^2\pi\mu\omega = nh$$

$$A^2 \equiv A_n^2 = \frac{nh}{\pi\mu\omega}$$

Substituting in Eq. (1),

$$E \equiv E_n = ½\,(\mu\omega^2)\left(\frac{nh}{\mu\pi\omega}\right) = nh\omega$$

PROBLEM–6

Find the eigenfunctions of a linear momentum operator, say $\left(\frac{h}{2\pi i}\frac{\partial}{\partial x}\right)$. Show that in the absence of any restriction eigenvalues of linear momentum form a continuous spectrum.

SOLUTION

Consider the equation

$$\frac{h}{2\pi i}\frac{\partial}{\partial x}\psi = \lambda.\psi$$

$$\frac{\partial\psi}{\partial x} = \frac{2\pi i}{h}\lambda.\psi$$

$$\psi = \exp\left(\frac{2\pi i}{h}\lambda\right)x$$

Let $k = \frac{2\pi}{h}\lambda$, then exp (ikx) is the eigenfunction with eigenvalue $kh/2\pi$. Since there is no restriction in the value of k, the eigenvalue of $\frac{h}{2\pi i}\frac{\partial}{\partial x}$ can form a continuous spectrum.

PROBLEM–7

Show that if ψ is an eigenfunction of A with the eigenvalue a, then it is also an eigenfunction of exp (A) with eigenvalue exp (a).

SOLUTION

$$\hat{A}\,\psi = a\psi$$

Hence $\quad \hat{A}^2\psi = \hat{A}^2(\hat{A}\psi) = \hat{A}(a\psi) = a\,(\hat{A}\psi) = a^2\psi$

Similarly $\quad \hat{A}\psi = \hat{A}\hat{A}\hat{A}\ldots\hat{A}\psi = a^n\psi$

Since exp $\quad A = 1 + A + A^2/2! + A^3/3! + \ldots$

$$= \sum_n \frac{1}{n!}A^n$$

We have

$$(\exp \hat{A})\,\psi = \left(\sum \frac{1}{n!}\hat{A}^n\right)\psi = \left(\sum_n \frac{1}{n!} a^n\right)\psi$$

$$= (\exp a).\,\psi$$

PROBLEM–8

Show that $[x^n, Px] = \frac{h}{[2\pi i]} nx^{n-1}$ where n is a positive integer.

SOLUTION

$$x^n p_x\,\psi = x^n \frac{h}{2\pi i}\frac{\partial\psi}{\partial x}$$

$$p_x\, x^n\psi = \frac{h}{2\pi i}\frac{\partial}{\partial x}x^n\psi = nx^{n-1}\frac{h}{2\pi i}\psi + x^n\frac{h}{2\pi i}\frac{\partial\psi}{\partial x}$$

hence

$$[x^n, p_x] = nx^{n-1}\frac{h}{2\pi i}$$

PROBLEM–9

Show that (a) the product of two hermitian operators is hermitian if they commute, and (b) the inverse of a hermitian operator is hermitian.

SOLUTION

(a) Let $\hat{A}$ and $\hat{B}$ be two hermitian operators that commute. Then the integral

$$\int \psi^* (\hat{A}\hat{B})\,\psi d\tau = \int \psi^*\hat{B}\hat{A}\psi d\tau = \int (\hat{A}\psi)\,\hat{B}^*\,\psi^*\,d\tau$$

$$= \int (\hat{B}\psi)^*\,(\hat{A}\psi)\,d\tau = \int \psi.\hat{A}^*.\,(\hat{B}\psi)^*\,d\tau = \int \psi.\,(\hat{A}\hat{B})^*\,\psi^*\,d\tau$$

(b) $\hat{A}\hat{A}^{-1} = \hat{A}^{-1}\hat{A} = \hat{I}$

Sine the unit operator $\hat{I}$ is hermitian, it follows from (a) that $\hat{A}^{-1}$ is hermitian.

PROBLEM–10

Find the normalized eigenfunction of the operator L_z.

SOLUTION

$\psi = A \exp\left(\frac{2\pi i}{h}\lambda\phi\right)$, where A is the normalisation factor. The normalisation integral must be unity, i.e.

$$\int_0^{2\pi} \psi^* \psi \, d\phi = 1$$

or
$$A^2 \int_0^{2\pi} \exp\left(-\frac{2\pi i}{h}\lambda\phi\right) . \exp\left(\frac{2\pi i}{h}\lambda . \phi\right) d\phi = 1$$

or
$$A^2 \int_0^{2\pi} d\phi = 1$$

or
$$A = \frac{1}{\sqrt{2\pi}}$$

The normalized eigenfunction is

$$\frac{1}{\sqrt{2\pi}} \exp\left(\frac{2\pi i}{h}\lambda\phi\right)$$

PROBLEM–11

Find the expression for the following operators :

(a) $\left(\frac{d}{dx} + x\right)^2$; (b) $\left(\frac{d}{dx} + \frac{1}{x}\right)^3$;

(c) $\left(x\frac{d}{dx}\right)^2$; (d) $\left(\frac{d}{dx}x\right)^2$.

SOLUTION

(a) Consider an arbitrary function $\psi(x)$ and operate upon it by the given operator. Thus.

$$\left(\frac{d}{dx} + x\right)^2 \psi(x) = \left(\frac{d}{dx} + x\right)\left(\frac{d}{dx} + x\right)\psi$$

$$= \left(\frac{d}{dx} + x\right)\left(\frac{d\psi}{dx} + x\psi\right)$$

$$= \frac{d^2\psi}{dx^2} + 2x\frac{d\psi}{dx} + x^2\psi + \psi$$

$$= \left(\frac{d^2}{dx^2} + 2x\frac{d}{dx} + x^2 + 1\right)\psi$$

Hence,

$$\left(\frac{d}{dx} + x\right)^2 = \frac{d^2}{dx^2} + 2x\frac{d}{dx} + x^2 + 1.$$

(b) $\left(\frac{d}{dx} + \frac{1}{x}\right)^3 = \frac{d^3}{dx^3} + \frac{3}{x}\frac{d^2}{dx^2}$

(c) $\left(x\frac{d}{dx}\right)^2 = x^2\frac{d^2}{dx^2}x\frac{d}{dx}$

(d) $\left(\frac{d}{dx}x\right)^2 = s^2\frac{d^2}{dx^2} + 3x\frac{d}{dx} + 1.$

PROBLEM–12

Show that

(a) $[\hat{A}, \hat{B}] = -[\hat{B}, \hat{A}]$

(b) $[\hat{A}^2, \hat{B}] = \hat{A}\,[\hat{A}, \hat{B}] + [\hat{A}, \hat{B}]\,\hat{A}$

(c) $[\hat{A}, [\hat{B}, \hat{C}]] = [[\hat{A}, \hat{B}], \hat{C}] + [\hat{B}, [\hat{A}, \hat{C}]]$

SOLUTION

(a) $[\hat{A}, \hat{B}] = (\hat{A}\hat{B} - \hat{B}\hat{A}) = -(\hat{B}\hat{A} - \hat{A}\hat{B}) = -[\hat{B}, \hat{A}]$

(b) R.H.S. $= \hat{A}[\hat{A}, \hat{B}] + [\hat{A}, \hat{B}]\,\hat{A}$

$= \hat{A}(\hat{A}\hat{B} - \hat{B}\hat{A}) + (\hat{A}\hat{B} - \hat{B}\hat{A})\,\hat{A}$

$= \hat{A}^2\hat{B} - \hat{A}\hat{B}\hat{A} + \hat{A}\hat{B}\hat{A} - \hat{B}\hat{A}^2$

$= \hat{A}^2\hat{B} - \hat{B}\hat{A}^2 = [\hat{A}^2, \hat{B}] =$ L.H.S.

(c) L.H.S. $= [\hat{A}, [\hat{B}, \hat{C}]]$

$= \hat{A}\,(\hat{B}\hat{C} - \hat{C}\hat{B}) - (\hat{B}\hat{C} - \hat{C}\hat{B})\,\hat{A}$

$= \hat{A}\hat{B}\hat{C} - \hat{A}\hat{C}\hat{B} - \hat{B}\hat{C}\hat{A} + \hat{C}\hat{B}\hat{A}$

R.H.S. $= [\hat{A}, \hat{B}], \hat{C}] + [\hat{B}, [\hat{A}, \hat{C}]]$

$= [(\hat{A}\hat{B} - \hat{B}\hat{A}), C] + [\hat{B},(\hat{A}\hat{C} - \hat{C}\hat{A})]$

$$= \hat{A}\hat{B}\hat{C} - \hat{B}\hat{A}\hat{C} - \hat{C}\hat{A}\hat{B} + \hat{C}\hat{B}\hat{A}$$
$$+ \hat{B}\hat{A}\hat{C} - \hat{B}\hat{C}\hat{A} - \hat{A}\hat{C}\hat{B} + \hat{C}\hat{A}\hat{B}$$
$$= \hat{A}\hat{B}\hat{C} - \hat{A}\hat{C}\hat{B} - \hat{B}\hat{C}\hat{A} + \hat{C}\hat{B}\hat{A} = \text{L.H.S.}$$

PROBLEM–13

(a) Evaluate the commutator $\left[x, \frac{d}{dx}\right]$

(b) If $\hat{A}$ and $\hat{B}$ are two operators such that $[\hat{A}, \hat{B}] = 1$, what is the value of $[\hat{A}, \hat{B}^2]$?

SOLUTION

(a) By definition,

$$\left[x, \frac{d}{dx}\right] = x\frac{d}{dx} - \frac{d}{dx}x$$

Operating on some arbitrary function $\psi(x)$ by the right-hand side expression,

$$\left(x\frac{d}{dx} - \frac{d}{dx}x\right)\psi = x\frac{d\psi}{dx} - \frac{d}{dx}(x\psi)$$

$$= x\frac{d\psi}{dx} - x\frac{d\psi}{dx} - \psi\frac{dx}{dx} = -\psi$$

Thus,
$$\left[x, \frac{d}{dx}\right] = -1$$

(b) $[\hat{A}, \hat{B}] = \hat{A}\hat{B} - \hat{B}\hat{A} = 1$

$[\hat{A}, \hat{B}^2] = \hat{A}\hat{B}^2 - \hat{B}^2\hat{A}$

Adding and subtracting $\hat{B}\hat{A}\hat{B}$ in the above expression,

$$[\hat{A}, \hat{B}^2] = \hat{A}\hat{B}^2 - \hat{B}^2\hat{A} + \hat{B}\hat{A}\hat{B} - \hat{B}\hat{A}\hat{B}$$
$$= (\hat{A}\hat{B}^2 - \hat{B}\hat{A}\hat{B}) + (\hat{B}\hat{A}\hat{B} - \hat{B}^2 A)$$
$$= (\hat{A}\hat{B} - \hat{B}\hat{A})\,\hat{B} + \hat{B}(\hat{A}\hat{B} - \hat{B}\hat{A})$$
$$= (1)\,\hat{B} + \hat{B}\,(1) = 2\hat{B}$$

PROBLEM–14

Show that

$$[x^n, p_x] = i\,hn\,x^{n-1},$$

where n is a positive integer.

SOLUTION

$$[x^n, p_x] = x^n p_x - p_x x^n$$

We know that the operators for the x-coordinate and the x-component of the linear momentum, p_x are :–

$$\hat{x} \rightarrow x$$

$$\hat{p}_x \rightarrow ih\frac{\partial}{\partial x}$$

$$[\hat{x}^n, \hat{p}_x^n] = x^n\left(-ih\frac{\partial}{\partial x}\right) - \left(-ih\frac{\partial}{\partial x}\right)x^n$$

$$= -ih\,x^n\frac{\partial}{\partial x} + ih\frac{\partial x^n}{\partial x}$$

Operating upon an arbitrary function $\psi(x)$ by this operator,

$$-ih\,x^n\frac{\partial\psi(x)}{\partial x} + ih\frac{\partial}{\partial x}(x^n\psi(x))$$

$$= -ihx^n\frac{\partial\psi}{\partial x} + ihx^n\frac{\partial\psi}{\partial x} + ihn\,x^{n-1}\,\psi = ihn\,x^{n-1}\,\psi$$

or, $$[\hat{x}^n, \hat{p}_x] = ihn\,x^{n-1}.$$

PROBLEM–15

Show that for operators $\hat{A}$, $\hat{B}$ and $\hat{C}$,

$$[\hat{A}\hat{B}, \hat{C}] = \hat{A}\,[\hat{B}, \hat{C}] + [\hat{A}, \hat{C}]\,\hat{B}$$

SOLUTION

$$[\hat{A}\hat{B}, \hat{C}] = \hat{A}\hat{B}\hat{C} - \hat{C}\hat{A}\hat{B}$$

Adding and subtracting $\hat{A}\hat{C}\hat{B}$ and rearranging

$$[\hat{A}\hat{B}, \hat{C}] = (\hat{A}\hat{B}\hat{C} - \hat{A}\hat{C}\hat{B}) - (\hat{C}\hat{A}\hat{B} - \hat{A}\hat{C}\hat{B})$$

$$= \hat{A}(\hat{B}\hat{C} - \hat{C}\hat{B}) - (\hat{C}\hat{A} - \hat{A}\hat{C})\hat{B}$$

$$= \hat{A}[\hat{B}, \hat{C}] - [\hat{C}, \hat{A}]\hat{B}$$

We know, however, that

$$[\hat{C}, \hat{A}] = -[\hat{A}, \hat{C}]$$

Hence, $[\hat{A}\hat{B}, \hat{C}] = A[\hat{B}, \hat{C}] + [\hat{A}, \hat{C}]\hat{B}$

PROBLEM–16

Prove that the quantum mechanical operators for the following observables are Hermitian : (a) linear momentum; (b) angular momentum; (c) energy.

SOLUTION

An operator $\hat{A}$ is said to be Hermitian if

$$\int \psi_i^*(\hat{A}\psi_j)d\tau = \int (\hat{A}\,\psi_i)^*\psi_j d\tau \qquad (1)$$

where ψ_i and ψ_j are eigenfunctions of $\hat{A}$.

(a) The operator for p_x, the x-component of the linear momentum, is

$$\hat{p}_x \rightarrow -ih\frac{d}{dx} \qquad (2)$$

In the x-direction $d\tau = dx$ and $-\infty < x < \infty$.

From the method of integration by parts we know that if u and v are functions of x, then

$$\int_a^b u\,dv = \Big[u\,v\Big]_a^b - \int_a^b v du \qquad (3)$$

Substituting Eq. (2) in Eq. (1) and using Eq. (3)

$$\int_{-\infty}^{\infty} \underset{u}{\underset{\uparrow}{\psi_i^*}} \underbrace{\left(-ih\frac{d}{dx}\right)\psi_j\,dx}_{dv} = -ih\Big[\underset{u}{\underset{\uparrow}{\psi_i^*}}\underset{v}{\underset{\uparrow}{\psi_j}}\Big]_{-\infty}^{\infty} + ih\int_{-inf}^{\infty} \underset{v}{\underset{\uparrow}{\psi_j}} \underbrace{\frac{d\psi_i^*}{dx}\,dx}_{du}$$

The first term in brackets on the R.H.S. of this equation is zero since by the fundamental postulate of quantum mechanics the wave functions ψ_i and ψ_j must vanish at infinity. The second term can be written as

$$\int_{-\infty}^{\infty} \psi_j \left(-ih\frac{d}{dx}\right)^* \psi_i^*\,dx$$

Hence, the operator for p_x is Hermitian. Similarly, the operators for p_y and p_z can be shown to be Hermitian.

(b) Proceed as in (a) above

$$\hat{L}_x \to -ih\left(y\frac{\partial}{\partial z} - z\frac{\partial}{\partial y}\right)$$

(c) For simplicity consider the Hamiltonian operator for the hydrogen atom, in Cartesian coordinates x, y, z.

$$\hat{H} = \hat{T} + \hat{V}$$

where the kinetic energy operator is

$$\hat{T} \to -\frac{h^2}{2m}\left(\frac{\partial^2}{\partial x^2} + \frac{\partial^2}{\partial y^2} + \frac{\partial^2}{\partial z^2}\right)$$

and the potential energy operator for the motion of the electron of charge $-e$ moving round the nucleus of charge $+Ze$ at a distance r from the nucleus is

$$V(r) \to -\frac{Ze}{r}.$$

Assuming that the wave functions are real, we have to show that

$$\int \psi_i \hat{H} \psi_j \, d\tau = \int \psi_j \hat{H} \psi_i \, d\tau \tag{4}$$

Let us consider the potential energy term of $\hat{H}$. Since $-Ze^2/r$ is just as multiplier, the order of terms under the integral sign is immaterial. Thus,

$$\int \psi_i \left(-\frac{Ze^2}{r}\right) \psi_j \, d\tau = \int \psi_j \left(-\frac{Ze^2}{r}\right) \psi_i d\tau = \int \left(-\frac{Ze^2}{r}\right) \psi_i \, \psi_j \, d\tau \tag{5}$$

Let us now consider the (d^2/dx^2) part of the kinetic energy operator. For it to be Hermitian we have to show that

$$\int_{-\infty}^{\infty} \psi_i \left(\frac{d^2}{dx^2}\right) \psi_j dx = \int_{-\infty}^{\infty} \psi_j \left(\frac{d^2}{dx^2}\right) \psi_i dx \tag{6}$$

Integrating the L.H.S. of Eq. (6) by parts,

$$\int_{-\infty}^{\infty} \underset{\substack{\uparrow \\ u}}{\psi_i} \underbrace{\left(\frac{d^2}{dx^2}\right) \psi_j dx}_{dv} = \left[\underset{\substack{\uparrow \\ u}}{\psi_i} \underset{\substack{\uparrow \\ v}}{\frac{d\psi_j}{dx}}\right]_{-\infty}^{\infty} - \int_{-\infty}^{\infty} \underset{\substack{\uparrow \\ v}}{\frac{d\psi_j}{dx}} \underbrace{\frac{d\psi_i}{dx} dx}_{du} \tag{7}$$

Again, integrating the R.H.S. of Eq. (6) by parts,

$$\int_{-\infty}^{\infty} \underset{u}{\psi_j} \underset{dv}{\underbrace{\left(\frac{d^2}{dx^2}\right)\psi_i\, dx}} = \left[\underset{u}{\psi_j}\, \underset{v}{\frac{d\psi_i}{dx}}\right]_{-\infty}^{\infty} - \int_{-\infty}^{\infty} \underset{v}{\frac{d\psi_i}{dx}}\, \underset{du}{\frac{d\psi_j}{dx}}\, dx \qquad (8)$$

Invoking the postulate of quantum mechanics according to which the wave function and its derivative must vanish at infinity, the first term in the brackets of both Eqs. (7) and (8) is zero. The second term is the same for the two equations since the integrals differ only in the order of multiplication of the two derivatives. Thus, the operator (d^2/dx^2) is Hermitian. Similarly, (d^2/dy^2) and (d^2/dz^2) can be shown to be Hermitian. (The factor $(-h^2/2m)$ in the operator for $\hat{T}$ is, of course, a constant multiplier). Hence, from Eqs (5), (7) and (8) both the potential energy and the kinetic energy components of $\hat{H}$ are Hermitian, so that $\hat{H}$ is Hermitian.

PROBLEM–17

Show that if ψ_i and ψ_j are eigenfunctions of the Hermitian operator $\hat{L}$ with eigenvalues λ_i and λ_j, respectively, then the eigenfunctions are orthogonal.

SOLUTION

For ψ_i and ψ_j to be orthogonal we have

$$\int \psi_i^* \psi_j \, d\tau = 0$$

By definition, a wave function is said to be an eigenfunction of an operator if the result of "operating" upon it by the operator is the wave function itself multiplied by a real constant called the "eigenvalue" of the operator. Thus, we have

$$\hat{L}\psi_i = \lambda_i \psi_i \qquad (1a)$$

$$\hat{L}\psi_j = \lambda_j \psi_j \qquad (1b)$$

Taking the complex-conjugate of (1a)

$$\hat{L}^*\psi_i^* = \lambda_i^*\psi_i^* \qquad (2)$$

Since λ_i is real (all experimental measurements carried out on a dynamical variable yield real numbers), therefore it is equal to its complex conjugate, i.e.,

$$\lambda_i^* = \lambda_i \tag{3}$$

Substituting in Eq. (2)

$$\hat{L}^*\psi_i^* = \lambda_i\psi_i^* \tag{4}$$

Multiplying Eq. (4) by ψ_j and integrating,

$$\int \psi_j\hat{L}^*\psi_i^* d\tau = \lambda_i \int \psi_j\psi_i^* \, d\tau \tag{5}$$

Similarly, multiplying Eq. (1b) by ψ_i*)and*integratingweget*

$$\int \psi_i^* \, \hat{L} \, \psi_j \, d\tau = \lambda_j \int \psi_i^*\psi_j \, d\tau \tag{6}$$

Since the operator $\hat{L}$ is Hermitian, the left hand sides of Eqs. (5) and (6) are equal, so that

$$\lambda_i \int \psi_j\psi_i^* d\tau = \lambda_j \int \psi_i^*\psi_j \, d\tau$$

or,

$$(\lambda_i - \lambda_j) \int \psi_i^*\psi_j d\tau = 0$$

Since

$$\lambda_i \neq \lambda_j,$$

$$\int \psi_i^*\psi_j d\tau = 0$$

Hence, ψ_i and ψ_j are orthogonal.

Note. A set of eigenfunctions ψ_i (i = 1, 2, 3, ..., n) is called orthonormal i.e., orthogonal and normalized, if

$$\int \psi_i^*\psi_j \, d\tau = \delta_{ij} = \{1, i = j above 0, i \neq j$$

where δ_{ij} is called the "Kronekcer delta".

PROBLEM–18

Two trains of travelling waves may be specified by the functions.

$$\psi_1 = \sin(ax + bt) \text{ and } \psi_2 = \sin(ax - bt)$$

(a) For each of these trains find the speed and direction of propagation and the wavelength. (Hint : Locate the nodes (where ψ = 0) for each train and see how they move.)

(b) show that

$$\psi_+ = \psi_1 + \psi_2 \text{ and } \psi_- = \psi_1 - \psi_2$$

represent standing waves. Locate the nodes in each wave.

SOLUTION

(a) $\psi_1 = 0$ when $ax + bt = \pi n$, where $n = 0, \pm 1, \pm 2, \ldots$

$$x = \frac{\pi n - bt}{a}$$

Each such nodal point is moving along the x axis in the negative direction with velocity $dx/dt = -b/a$. Similarly for ψ_2, $dx/dt = +b/a$; this wave propagates along the x axis in the positive direction. The wavelength in each case will be twice the distance between adjacent nodes (see Fig. 3).

$$\lambda = 2(x_{n+1} - x_n) = \frac{2\pi}{a}$$

(b) $\psi_+ = \sin(ax + bt) + \sin(ax - bt)$

$= \sin ax \cos bt + \cos ax \sin bt$

$+ \sin ax \cos bt - \cos ax \sin bt$

$= 2 \sin ax \cos bt$

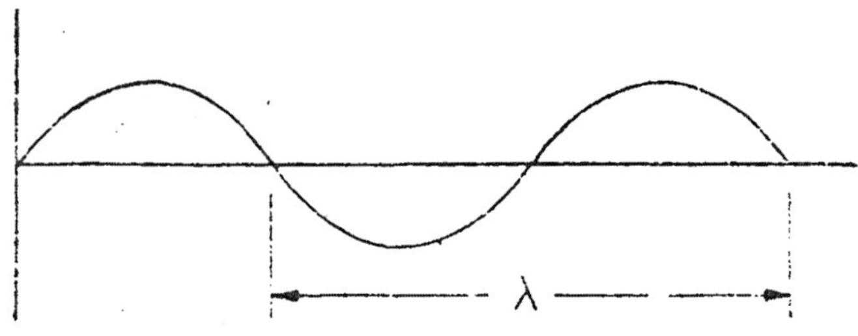

Fig. 3

This function will have a node wherever $x = \pi n/a$. These nodes do not move with time and the function therefore represents a standing wave. Similarly

$$\psi_- = 2 \cos a x \sin bt$$

There is a node wherever

$$x = \frac{\pi(n + \frac{1}{2})}{a}$$

PROBLEM–19

Which of the following are given functions of the operator d^2/dx^2? Give the eigen value where appropriate.

(a) $\sin x$ (b) $\cos x$

(c) $\sin^2 x$ (d) e^{2x}

(d) e^{ix} (f) $\sin nx$

SOLUTION

(a) $Qf = \frac{d^2}{dx^2} \sin x = -\sin x = -f$; the eigen value is -1.

(b) $Qf = \frac{d^2}{dx^2} \cos x = -\cos x = -f$; the eigen value is -1.

(c) $Qf = \frac{d^2}{dx^2} \sin^2 x = 2 - 4 \sin^2 x$; it is not an eigen function.

(d) $Qf = \frac{d^2}{dx^2} e^{2x} = 4f$: the eigen value is $+4$.

(e) $Qf = \frac{d^2}{dx^2} e^{1x} = -f$; the eigen value is -1.

(f) $Qf = \frac{d^2}{dx^2} \sin nx = -n^2 f$, the eigen value is $-n^2$.

PROBLEM–20

Examine the solutions of $Qf = \pm cf$, where $Q = \frac{d^2}{dx^2}$, and find the eigen functions for the space $-\infty < x < \infty$. (c is a positive constant.)

SOLUTION

First, consider the plus sign. The solution of

$$\frac{d^2 f}{dx^2} = cf$$

is

$$f = ac\, \sqrt{c.x.} \pm Be^{-} \sqrt{c.x.}.$$

This function diverges as $x \to \pm\infty$, so it is not a suitable eigen-function. However, if A were zero, the function

$$f = Be^{-}\sqrt{c.x.}$$

would be well behaved in the positive half-space ($x > 0$). For $B = 0$, the function

$$f = Ae\sqrt{c.x.}$$

would be well behaved in the negative half-space ($x < 0$).

To obtain well-behaved eigenfunctions for all of space, we must consider eigen-values $-c$. Then,

$$\frac{d^2f}{dx^2} = -cf$$

and the solution of this equation is

$$f = D\sin\sqrt{c.x} + E\cos\sqrt{c.x.}$$

This function is well behaved for all x, and an eigenfunction of Q exists for all eigen-values $-c$. For this situation we obtain a continuum of possible eigenvalues, whereas in many physical problems the eigenvalues form a discrete set of numbers.

PROBLEM–21

The time-independent one-dimensional Schrodinger equation is

$$\left(\frac{-h^2}{8\pi^2 m}\right)\left(\frac{\delta^2\psi}{\partial x^2}\right) + V(x)\,\psi = E\psi$$

where V is the potential energy. It is well known that only differences in V are physically significant; nothing is essentially changed if a constant is added to V. Show how the Schrodinger equation is affected by such a change in V and why the effect is inessential.

SOLUTION

Let $$V' = V + C, \text{ or } V = V' - C$$

$$-\frac{h^2}{8\pi^2 m}\frac{d^2\psi}{dx^2} + (V' - C)\,\phi = E\psi$$

$$-\frac{h^2}{8\pi^2 m} + V'\psi = (E + C)\psi$$

Thus, the equation can equally well be written in terms of V' instead of V but each eigenvalue of energy is then increased by the same constant C. A list of energies is defined relative to a certain choice for the origin of V.

PROBLEM–22

A postulate of quantum mechanics is that the operator corresponding to energy is $-(h/2\pi i)\ (\delta/\partial t)$, where t is time.

(a) What must be the time dependence of an eigenfunction of energy? (In other words, what kind of function is a solution of the equation $-(h/2\pi i)\ (\delta\psi/\partial t) = E\psi$?)

(b) When ψ is an eigenfunction of energy, now does the probability density $|\psi|^2$ depend on time? How does this result help to explain why eigenfunctions of energy are so important, especially in chemistry?

SOLUTION

(a)
$$-\frac{h}{2\pi i}\frac{\partial\psi}{\delta t} = E\psi$$

$$-\frac{h}{2\pi i}\frac{\partial \ln\psi}{\partial t} = E$$

$$\ln\psi = \frac{-2\pi Et}{h} + \text{constant}$$

$$\psi = \psi_0\, e^{-2\pi eEt/h}$$

(b) $|\psi|^2 = |\psi_0|^2\, e^{-2\pi iEt/h^2} = |\psi_0|^2$

since $|e^{i\phi}| = 1$ for any real ϕ. Thus $|\psi|^2$ does not depend on time. An eigenfunction of energy corresponds to a stable probability distribution. Chemistry is concerned largely with long-lived, stable molecules, in which the electron distribution remains constant in time. (Note that ψ_0, though constant with respect to time, may depend on x, y, and z.)

PROBLEM–23

Show that if ψ_i, and ψ_j are eigenfunction of the operator $\hat{L}$ with eigenvalues λ_i and λ_j, respectively ($\lambda_i \neq \lambda_j$), then

$$\int \psi_i^* \hat{L}\psi_j d\tau = 0$$

SOLUTION

$$\hat{L}\,\psi_i = \lambda_i\psi_i$$

$$\hat{L}\psi_j = \lambda_j\psi_j$$

Thus,
$$\int \psi_i^* \hat{L}\psi_j \, d\tau = \int \psi_i^* \lambda_j \psi_j \, d\tau$$

The eigenvalue, λ_j being a constant, can be taken out of the integral, giving

$$\int \psi_i^* \hat{L}\psi_j \, d\tau = \lambda_j \int \psi_i^* \psi_j \, d\tau$$

But by the orthogonality condition,

$$\int \psi_i^* \psi_j \, d\tau = 0 \quad (i \neq j)$$

Hence,
$$\int \psi_i^* \hat{L}\psi_j \, d\tau = 0$$

PROBLEM–24

Show that if λ is an eigenvalue of an operator $\hat{L}$, then λ^n (where n is an integer) is an eigenvalue of the operator $\hat{L}^n$.

SOLUTION

$$\hat{L}\psi_i = \lambda_i\psi_i$$

$$L^2\psi_i = \hat{L}\,(\hat{L}\psi_i) = \hat{L}(\lambda_i\psi_i)$$

$$= \lambda_i\,(\hat{L}\psi_i) = \lambda_i\lambda_i\psi_i$$

$$= \lambda_i^2\psi_i$$

Generalizing to higher powers of $\hat{L}$,

$$\hat{L}^n\psi_i = \lambda_i^n\psi_i$$

Hence, λ_i^n is the eigenvalue of $\hat{L}^n$.

PROBLEM–25

The spin of an electron can have two orientations in space : upwards (↑) and downwards (↓). It is customary to associate a spin wave function α with the upward orientation and a spin wave function β with the downward orientation. How will you express mathematically the fact that α and β are normalized and orthogonal?

SOLUTION

Let s be the spin variable. The spin wave functions are functions of s. Their normalization and orthogonality conditions are, respectively:

$$\int \alpha^*(s)\alpha(s)ds = \int \beta^*(s)\beta(s)ds = 1$$

$$\int \alpha^*(s)\beta(s)ds = \int \beta^*(s)\alpha(s)ds = 0$$

the integration being carried over the spin space.

PROBLEM–26

If an arbitrary function ψ can be expressed as a linear combination of a complete set of orthonormal functions ψ_i :

$$\psi = \sum_{i=1}^{n} c_i\psi_i = c_1\psi_1 + c_2\psi_2 + c_3\psi_3 + \ldots + c_n\psi_n$$

derive an expression for the coefficients in the above expansion.

SOLUTION

$$\psi = c_1\psi_1 + c_2\psi_2 + c_3\psi_3 \ldots + c_j\psi_j + \ldots + c_n\psi_n$$

Multiplying both sides of the above equation by ψ_j^* (where j varies from 1 and n), we get

$$\psi_j^* \psi = c_1\psi_j^*\psi_1 + c_2\psi_j^*\psi_2 + \ldots + c_j\psi_j^*\psi_j + \ldots c_n\psi_j^*\psi_n.$$

Integrating this expression over all coordinate space,

$$\int \psi_j^* \psi \, d\tau = c_1 \int \psi_j^*\psi_1 d\tau + c_2 \int \psi_j^*\psi_2 d\tau + \ldots$$

$$+ c_j \int \psi_j^*\psi_j d\tau + \ldots + c_n \int \psi_j^*\psi_n d\tau$$

Since the ψ's are orthogonal and noramlized,

$$\int \psi_j^* \psi \, d\tau = c_1 \times (0) + c_2 \times (0) + \ldots + c_j \times (1) + \ldots c_n \times (0)$$

$$= c_j$$

Thus, $$c_j = \int \psi_j^* \psi \, d\tau$$

PROBLEM–27

Let E_0 be the ground-state energy of a system described by the Hamiltonian $\hat{H}$ and the ground-state wave function ψ_0. Show that the total energy of the system can be determined exactly.

SOLUTION

The Schrodinger equation for the system is

$$\hat{H}\psi_0 = E_0\psi_0$$

$$<E> = \int \psi_0\hat{H}\psi_0 d\tau = \int \psi_0 E_0 \psi_0 d\tau = E_0 \int \psi_0\psi_0 d\tau = E_0$$

($\because$ ψ_0 is normalized)

Again,

$$<E^2> = \int \psi_0\hat{H}^2\psi_0 d\tau = \int \psi_0\hat{H}(\hat{H}\psi_0)\, d\tau = \int \psi_0\hat{H}(E_0\psi_0)\, d\tau$$

$$= E_0 \int \psi_0\hat{H}\psi_0\, d\tau = E_0^2$$

Since uncertainty in energy is defined as

$$\Delta E = \left[<E^2> - <E>^2\right]^{1/2}$$

therefore, $\Delta E = E_0^2 - (E_0)^2 = 0.$

Hence, apart from the limitations imposed by Δt, the uncertainty in time, the total energy of a system can be determined exactly.

PROBLEM–28

Prove that the average value of the square of a Hermitian operator is positive.

SOLUTION

Let $\hat{L}$ be a Hermitian operator. The average value of $\hat{L}$ is defined as :

$$<\hat{L}> = \frac{\int \psi^*\hat{L}^2\psi d\tau}{\int \psi^*\psi d\tau}$$

In this expression the denominator is, of course, positive. So, we have to prove that the numerator is positive. Since $\hat{L}$ is Hermitian,

$$\int \psi^*\hat{L}^2\psi d\tau = \int \psi^*\hat{L}\hat{L}\psi d\tau = \int (\hat{L}\psi)\,(\hat{L}^*\psi^*)\, d\tau$$

$$= \int |\hat{L}\psi|^2 d\tau > 0$$

PROBLEM–29

Show that for a particle of mass m, moving under the three-dimensional potential, V(x, y, z)

$$\int \psi_j^* \left(\frac{\partial}{\partial x}\right) \psi_k d\tau = \frac{m}{h^2}(E_j - E_k) \int \psi_j^* \, x \, \psi_k d\tau$$

where ψ_j and ψ_k are the eigenfunctions of the Hamiltonian belonging to the eigenvalues E_j and E_k, respectively.

SOLUTION

$$\hat{H} = -\frac{h^2}{2m}\left(\frac{\partial^2}{\partial x^2} + \frac{\partial^2}{\partial y^2} + \frac{\partial^2}{\partial z^2}\right) + V(x, y, z) \quad (1)$$

We will make use of the fact that the only part of $\hat{H}$ that does not commute with x is $\partial^2/\partial x^2$:

$$\left[x, \frac{\partial^2}{\partial x^2}\right] \neq 0; \left[x, \frac{\partial^2}{\partial y^2}\right] = \left[x, \frac{\partial^2}{\partial z^2}\right] = [x, V] = 0 \quad (2)$$

$$[\hat{H}, \hat{x}] = \hat{H}\,\hat{x} - \hat{x}\,\hat{H}.$$

Operating on the function ψ_k,

$$(\hat{H}x - x\hat{H})\,\psi_k = -\frac{h^2}{2m}\left(\frac{\partial^2}{\partial x^2}x - x\frac{\partial^2}{\partial x^2}\right)\psi_k$$

$$= -\frac{h^2}{2m}\left[\frac{\partial^2(x\psi_k)}{\partial x^2} - x\frac{\partial^2\psi_k}{\partial x^2}\right]$$

$$= -\frac{h^2}{2m}\left[\frac{\partial}{\partial x}\left\{x\frac{\partial\psi_k}{\partial x} + \psi_k\right\} - x\frac{\partial^2\psi_k}{\partial x^2}\right]$$

$$= -\frac{h^2}{2m}\left[x\frac{\partial^2\psi_k}{\partial x^2} + 2\frac{\partial\psi_k}{\partial x} - x\frac{\partial^2\psi_k}{\partial x^2}\right]$$

$$= -\frac{h^2}{m}\frac{\partial\psi_k}{\partial x} \quad (3)$$

Multiplying both sides of Eq. (3) on the left by ψ_j^* and integrating,

$$\int \psi_j^*(\hat{H}x - x\hat{H})\psi_k \, d\tau = -\frac{h^2}{m}\int \psi_{j*}\frac{\partial\psi_k}{\partial x}\, d\tau \quad (4)$$

Because of the Hermitian character of $\hat{H}$,

$$\int \psi_j^*(\hat{H}x)\psi_k d\tau = \int x\psi_k \hat{H}^* \psi_j^* d\tau = E_j \int x\psi_k\psi_j^* d\tau$$

$$= \int \psi_j^* \, x \, \psi_k \, d\tau$$

Similarly, $$\int \psi_j^* \, (x\hat{H})\psi_k d\tau = E_k \int \psi_j^* \, x \, \psi_k \, d\tau$$

Substituting in Eq. (4),

$$(E_j - E_k) \int \psi_j^* \, x \, \psi_k \, d\tau = -\frac{h^2}{m} \int \psi_j^* \frac{\partial \psi_k}{\partial x} \cdot d\tau$$

Thus, $$\int \psi_j^* \left(\frac{\partial}{\partial x} \right) \psi_k d\tau = -\frac{m}{h^2} (E_j - E_k) \int \psi_j^* \, x \, \psi_k d\tau$$

PROBLEM–30

Find the expression for the following operators.

(a) $\left(A - \dfrac{d}{dA} \right)$ (b) $\left(\dfrac{d}{dA} A \right)^2$

(c) $\left(\dfrac{d}{dA} + A \right)^2$ (d) $\left(\dfrac{d}{dA} + 1 \right)\left(\dfrac{d}{dA} - 1 \right)$

SOLUTION

(a) Consider an arbitrary function ψ (A) and operate upon it by the given operator. Thus,

$$\left(A \frac{d}{dA} \right)^2 \psi(A) = \left(A \frac{d}{dA} \right)\left(A \frac{d}{dA} \right) \psi$$

$$= \left(A \frac{d}{dA} \right)\left(A \frac{d\psi}{dA} \right)$$

$$= A^2 \frac{d^2\psi}{dA^2} + A \frac{d\psi}{dA}$$

$$= \left(A^2 \frac{d^2}{dA^2} + A \frac{d}{dA} \right) \psi$$

Hence $$\left(A \frac{d}{dA} \right)^2 = A^2 \frac{d^2}{dA^2} + A \frac{d}{dA}$$

(b) Consider an arbitrary function ψ (A) and operate upon it by the given operator, Thus,

$$\left(\frac{d}{dA}A\right)^2\psi(A) = \left(\frac{d}{dA}A\right)\left(\frac{d}{dA}A\right)\psi$$

$$= A^2\frac{d^2\psi}{dA^2} + 3A\frac{d\psi}{dA} + \psi$$

$$= \left(A^2\frac{d^2}{dA^2} + 3A\frac{d}{dA} + 1\right)\psi$$

Hence $$\left(\frac{d}{dA}A\right)^2\psi(A) = A^2\frac{d^2}{dA^2} + 3A\frac{d}{dA} + 1$$

(c) Consider an arbitrary function (ψ) A and operate upon it by the given operator. Thus,

$$\left(\frac{d}{dA} + A\right)^2\psi(A) = \left(\frac{d}{dA} + A\right)\left(\frac{d}{dA} + A\right)\psi$$

$$= \left(\frac{d}{dA} + A\right)\left(\frac{d\psi}{dA} + A\psi\right)$$

$$= \frac{d^2\psi}{dA^2} + 2A\frac{d\psi}{dA} + A^2\psi + \psi$$

$$= \left(\frac{d^2}{dA^2} + 2A\frac{d}{dA} + A^2 + 1\right)\psi$$

Hence $$\left(\frac{d}{dA} + A\right)^2 = \left(\frac{d^2}{dA^2} + 2A\frac{d}{dA} + A^2 + 1\right)$$

(d) $$\left(\frac{d}{dA} + A\right)\left(\frac{d}{dA} - A\right)\psi(A)$$

$$= \left(\frac{d}{dA} + A\right)\left(\frac{d}{dA} - A\right)\psi$$

$$= \left(\frac{d}{dA} + A\right)\left(\frac{d\psi}{dA} - A\psi\right)$$

$$= \frac{d^2\psi}{dA^2} - \frac{dA}{dA}\psi - A\frac{d\psi}{dA} + A\frac{d\psi}{dA} - A^2\psi$$

$$= \frac{d^2\psi}{dA^2} - \psi - A^2\psi$$

$$= \left(\frac{d^2}{dA^2} - 1 - A^2\right)\psi$$

Hence $$\left(\frac{d}{dA} + A\right)\left(\frac{d}{dA} - A\right) = \left(\frac{d^2}{dA^2} - 1 - A^2\right)$$

PROBLEM–31

Prove that $[\hat{X}, \hat{Y}] = -[\hat{Y}, \hat{X}]$

SOLUTION

$$[\hat{X}\hat{Y}] = (\hat{X}\hat{Y} - \hat{Y}\hat{X})$$

$$= -(\hat{Y}\hat{X} - \hat{X}\hat{Y})$$

$$= -[\hat{Y}, \hat{X}]$$

PROBLEM–32

Determine the eigenfunctions and eigenvalues of (a) the linear momentum operator, $\hat{p}_z$; (b) the angular momentum operator, $\hat{L}_z$.

SOLUTION

(a) The eigenvalues equation is

$$\hat{p}_z\psi = \lambda\psi \tag{1}$$

$$p_z \rightarrow -ih\frac{\partial}{\partial z}$$

$$-ih\frac{\partial}{\partial z}\psi = \lambda\psi \tag{2}$$

$$\frac{\partial\psi}{\partial z} - \frac{i\lambda}{h}\psi = 0$$

$$\frac{\partial\psi}{\partial z} - im\,\psi = 0; \left(\text{where } m = \frac{\lambda}{h}\right) \tag{3}$$

Eq. (3) has the solution $\psi = exp\,(imz)$, as can be easily verified. Substituting this value of ψ in Eq. (2),

$$(-ih)\,(im)\exp(imz) = \lambda\exp(imz)$$

or, $\lambda = mh$, where m has any value.

Since there is no restriction on the values, of m, the eigenvalues of p_z are not quantized, i.e., they are continuous.

(b) The eigenvalue equations is

$$\hat{L}_z \psi = \lambda\psi \qquad (1)$$

L_z is the component of the angular momentum along the z-axis.

$$\hat{L}_z \to -ih\frac{\partial}{\partial\phi}$$

where ϕ is the angle the angular momentum vector makes with the z-axis.

$$-ih\frac{\partial}{\partial\phi}\psi = \lambda\psi \qquad (2)$$

$$\frac{\partial\psi}{\partial\phi} - \frac{i\lambda}{h}\psi = 0$$

$$\frac{\partial\psi}{\partial\phi} - im\,\psi = 0\left(\text{where } m = \frac{\lambda}{h}\right) \qquad (3)$$

whose solution is

$$\psi = exp\ (im\phi)$$

Since, by definition, the wave functions ψ must be single-valued for any ϕ $(0 < \phi < 2\pi)$

$$\psi(\phi) = \psi(\phi + 2\pi),$$

$$e^{im\phi} = e^{im(\phi + 2\pi)}$$

$$e^{2\pi im} = 1$$

or, $$\cos\ (2\pi m) + i\ \sin\ (2\pi m = 1$$

which is true only if

$$m = 0, \pm1, \pm2, \pm3,...$$

Hence, $$\lambda = mh.$$

Thus, the eigenvalues of $\hat{L}_z$ are either zero or an integral multiple of the Dirac's constant, h. So, unlike $\hat{p}_z$, the eigenvalues of $\hat{L}_z$ form a discrete set, i.e., they are quantized. Recall that this was one of the postulates of the Bohr theory of hydrogen atom.

PROBLEM–33

(a) Show that the function ψ(x) = exp (ikx) is an eigenfunction of p_x.

(b) Show that the functions sin mϕ or cos mϕ are not eigenfunctions of $\hat{L}_z$ but exp (imϕ) is an eigenfunction of this operator.

SOLUTION

(a) $\hat{p}_x \rightarrow -ih\frac{\partial}{\partial x}$

$$\hat{p}_x\psi(x) = i\,ih\frac{\partial}{\partial x}(e^{ikx}) = -ih\,(ik)e^{ikx} = kh\,e^{ikx}.$$

Hence, *exp* (*ikx*) is an eigenfunction of $\hat{p}_x$ with the eigenvalue *kh*, where k can have any value. the reader can easily prove part (b).

PROBLEM–34

Show that the wavelength of the wave represented by the function ψ(x) = exp (ikx) is that given by the de-Broglie relation.

SOLUTION

$$\psi(x) = e^{ikx} \qquad (1)$$

Changing x to $x + \lambda$, where λ is the de=Broglie wave length,

$$\psi(x+\lambda) = e^{ik(x+\lambda)} = e^{ikx} + e^{ik\lambda}$$

Since

$$\lambda = \frac{h}{p}$$

$$= \frac{h}{kh} = \frac{2\pi}{k}$$

Hence,

$$\psi\,(x+\lambda) = e^{ikx}\,e^{2\pi i}$$

$$= e^{ikx}$$

$$= \psi(x) \qquad (2)$$

That is, the wave function repeats itself after an interval of λ.

PROBLEM–35

If ψ_p and ψ_q represents the eigen functions of the operator A with eigen values λ_p and λ_q respectively ($\lambda_p \neq \lambda_q$), the prove that

$$\int \psi_p{*}\, \hat{A}\psi_q\, d\tau = 0$$

SOLUTION

But
$$\hat{A}\,\psi_p = \lambda_p\,\psi_p$$

$$\hat{A}\,\psi_q = \lambda_q\,\psi_q$$

$$\int \psi_p{*}\,\hat{A}\,\psi_q\,d\tau = \lambda_q \int \psi_p{*}\,\psi_q\,d\tau$$

As the eigen value λ_q is a constant, it can be taken out of the integral giving

$$\int \psi_p{*}\,\hat{A}\,\psi_q\,d\tau = \lambda_q \int \psi_p{*}\,\lambda_q\,d\tau$$

We know that the orthogonality condition is as follows :

$$\int \psi_p{*}\,\psi_q\,d\tau = 0 \;\; (p \neq q)$$

Thus,
$$\int \psi_p{*}\,\hat{A}\,\psi_q\,d\tau = 0$$

PROBLEM–36

What would happen to the time-independent three-dimensional Schrodinger equation if the potential energy is put equal to the sum of three terms each of which is a function of one variable only?

SOLUTION

The three-dimensional Schrodinger's wave equation may be put as follows :

$$\nabla^2\,\psi(r) + \frac{2m}{h^2}\,[E - V(r)]\psi(r) = 0 \qquad (1)$$

where
$$\nabla^2 = \frac{\partial^2}{\partial x^2} + \frac{\partial^2}{\partial y^2} + \frac{\partial^2}{\partial z^2}$$

$$V(r) = V(x, y, z) = V_1\,(x) + V_2\,(y) + V_3\,(z)$$

if $x \to x_1, y \to x_2, z \to x_3$, the we obtain

$$\nabla^2 + \frac{\partial^2}{\partial x_1^2} + \frac{\partial^2}{\partial x_2^2} + \frac{\partial^2}{\partial x_3^2} \qquad (2)$$

$$V(r) = V\,(x_1, x_2, x_3)$$

$$= V_1\,(x_1) + V_2\,(x_2) + V_3\,(x_3) \qquad (3)$$

Further, it may be assumed that

$$E + E_1 + E_2 + E_3 \tag{4}$$

and

$$\psi(r)\ \psi_1(x_1)\ \psi_2(x_2)\ \psi_3(x_3) \tag{5}$$

If we substitute Eqs. (2), (3), (4), and (5) in Eq. (1), we get

$$\left(\frac{\partial^2}{\partial x_1^2} + \frac{\partial^2}{\partial x_2^2} + \frac{\partial^2}{\partial x_3^2}\right)[\psi_1(x_1)\psi_2(x_2)\psi_3(x_2)]$$

$$\frac{2m}{h^2}\left[(E_1 + E_2 + E_3) - \left\{V_1(x_1) + V_2(x_2) + V_3(x_2)\right\}\right]$$

$$\psi_1(x_1)\psi_2(x_2)\psi_3(x_3) = 0$$

If the above equation is simplified and divided by $\psi_1(x_1)\psi_2(x_2)\psi_3(x_3)$, we obtain

$$\sum_{i=1}^{3}\left[\frac{1}{\psi_1}\frac{d^2\psi}{dx_i^2} - \frac{2m\, V_i(x_i)}{h^2}\right] = \frac{-2m}{h^2}\sum_{i=1}^{3} E_i$$

or

$$\frac{1}{\psi_i}\frac{d^2\psi}{dx^2} - \frac{2m\, V_i(x_i)}{h^2} = \frac{-2m}{h^2} E_i \ (i = 1, 2, 3)$$

or

$$\frac{d^2\psi}{d\,x_i^2} + \frac{2m}{h^2}\left[E_i - V_i(x_i)\right]\psi_i = 0 \ (i = 1, 2, 3) \tag{6}$$

Thus, it can be concluded that as a result of partitioning of the potential energy, the three-dimensional Schrodinger's equation (1) gets splitted into three one-dimensional equations (6).

PROBLEM-37

Show that if ϕ_1 and ϕ_2 are the wave functions of a degenerate energy E, then any linear combination of the two functions, viz., $c_1\phi_1 + c_2\phi_2$, where c_1 and c_2 are constants, is also an eigenfunction of E.

SOLUTION

$$\hat{H}\phi_1 = E\phi_1$$

$$\hat{H}\phi_2 = E\phi_2$$

$$\phi = c_1\phi_1 + c_2\phi_2$$

Thus,

$$\hat{H}\phi = \hat{H}(c_1\phi_1 + c_2\phi_2) = c_1\hat{H}\phi_1 + c_2\hat{H}\phi_2$$

$$= c_1 E\phi_1 + c_2 E\phi_2 = E(c_1\phi_1 + c_2\phi_2) = E\phi.$$

Hence, ϕ is an eigenfunction of $\hat{H}$ with energy E as the eigenvalue.

PROBLEM–38

Show that the energy levels of bound states in one-dimensional quantum mechanical systems are always non-degenerate.

SOLUTION

Suppose that the energy levels are degenerate. Suppose further that $\psi_1(x)$ and $\psi_2(x)$ are the two linear independent eigenfunctions belonging to the same energy eigenvalue E. ψ_1 and ψ_2 obviously satisfy Schrodinger equation

$$-\frac{h^2}{2m}\frac{d^2\psi_1(x)}{dx^2} + V(x)\psi_1(x) = E\psi_1(x) \tag{1}$$

$$-\frac{h^2}{2m}\frac{d^2\psi_2(x)}{dx^2} + V(x)\psi_2(x) = E\psi_2(x) \tag{2}$$

Using the abbreviated notation

$$\psi' = \frac{d\psi}{dx};\ \psi'' = \frac{d^2\psi}{dx^2}.$$

Eqs. (1) and (2) can be written as

$$\psi_1'' + \frac{2m}{h^2}(E - V)\psi_1 = 0 \tag{3}$$

$$\psi_2'' + \frac{2m}{h^2}(E - V)\psi_2 = 0 \tag{4}$$

Thus, $$\frac{\psi_1''}{\psi_1} = \frac{\psi_2''}{\psi_2} = \frac{2m}{h^2}(V - E)$$

or,

$$\psi_1''\psi_2 - \psi_2''\psi_1 = (\psi_1'\psi_2)' - (\psi_2'\psi_1)' = 0$$

Integration of this equation yields

$$\psi_1'\psi_2 - \psi_2'\psi_1 = C$$

where C is a constant.

Since for bound states ψ_1 and ψ_2 must be equal at infinity.

$$C = 0$$

or
$$\frac{\psi_1'}{\psi_1} = \frac{\psi_2'}{\psi_2}$$

Further integration of this equation gives

$$\ln \psi_1 = \ln \psi_2 + \ln C$$

or
$$\psi_1 = C\psi_2$$

whence it follows that ψ_1 and ψ_2 are not linearly independent. But we have assumed above that ψ_1 and ψ_2 are linearly independent. Hence, the energy levels of bound states in one-dimensional systems are always non-degenerate.

PROBLEM–39

What is the physical meaning of the fact that if in the time- dependent Schrodinger equation (TDSE) the potential does not depend explicitly on the time, the Schrodinger equation can be written with a plus or a minus sign in front of the time derivative :

$$\pm ih \frac{\partial \psi}{\partial t} = \left[-\frac{h^2}{2m} \nabla^2 + V(r)\right] \psi (r, t).$$

SOLUTION

Using plus sign in front of the time derivative,

$$\psi_n (x, t) = \psi_n (x) \exp (-iE_n t/h).$$

Probability density is:

$$P (x, t) = |\psi_n (x, t)|^2 = \psi^* \psi = |\psi_n (x)|^2$$

Using minus sign in front of the time-derivative,

$$\psi_n (x, t) = \psi_n (x) \exp (iE_n t/h)$$

$$\text{or } P (x, t) = \psi^* \psi = |\psi|^2 = |\psi_n (x)|^2.$$

Thus, both solutions lead to the same probability density.

PROBLEM–40

Show that in the time-independent one-dimensional Schrodinger equation if the wave function $\phi(x)$ of the particle moving under the potential V(x0 = 1/2 kx^2 (where k is a constant), is changed to ψ(nx), where n is a constant, the average kinetic energy increases by a factor of n^2 and the average potential energy decreases by a factor of n^2.

SOLUTION

The operator for kinetic energy is

$$\hat{T} \rightarrow -\frac{h^2}{2m}\frac{d^2}{dx^2}$$

$$<\hat{T}> = \int \psi^*(x)\,\hat{T}\,\psi(x)dx = \int \psi^*(x)\left[-\frac{h^2}{2m}\frac{d^2}{dx^2}\right]\psi(x)\,dx$$

Similarly,

$$<V> = \int \psi^*(x)\,[\tfrac{1}{2}kx^2]\,\psi(x)dx$$

As $\qquad \psi(x) \rightarrow \psi(nx),$

$$<\hat{T}> = \int \psi^*(nx)\left[-\frac{h^2}{2m}\frac{d^2}{dx^2}\right]\psi(nx)dx$$

$$= n^2\int \psi^*(nx)\left(-\frac{h^2}{2m}\right)\frac{d^2\psi(nx)}{d(nx)^2}\,dx = n^2\int \psi(x)\left[-\frac{h^2}{2m}\frac{d^2}{dx^2}\right]\psi(x)dx$$

$$<V> = \int \psi^*(nx)\,[\tfrac{1}{2}kx^2]\,\psi(nx)dx$$

$$= \frac{1}{n^2}\int \psi^*(nx)\left[\tfrac{1}{2}k\,(kx)^2\right]\psi(nx)dx$$

$$\frac{1}{n^2}\int \psi(x)\left[\tfrac{1}{2}kx^2\right]\psi(x)dx.$$

Hence, the result.

PROBLEM–41

Show for a bound state if the normalisation of the wave function is to be constant in time, it is necessary and sufficient that the Hamiltonian be hermitian.

SOLUTION

The normalisation of a wave function would be constant provided

$$N = \int |\psi(\vec{r},t)|^2\,\vec{dr} \text{ is a constant.}$$

or
$$\frac{dN}{dt} = 0$$

i.e.
$$\int \left(\psi^* \frac{d\psi}{dt} + \psi \frac{d\psi}{dt} \right) d\vec{r} = 0$$

Integration is done over the entire configuration space.

$$\int [\psi^* H\psi - (H\psi)^* \psi]\, d\vec{r} = 0$$

As ψ is any arbitrary quantum mechanical state, the above equation is true for any ψ. Hence H must be hermitian.

PROBLEM–42

Show by the partial integration of the following equation

$$\langle p_x \rangle = \int \psi^* \left(-ih \frac{\partial \psi}{\partial x} \right) d\tau$$

p_x is real.

SOLUTION

$$\langle p_x \rangle = -ih \iint dy\, dz \left[\psi^* \psi \Big|_{-\infty}^{+\infty} - \int_{-\infty}^{+\infty} \psi \frac{\partial \psi^*}{dx} \right]$$

$$= ih \int \psi \frac{\partial \psi^*}{\partial x} d\tau$$

$$= \langle p_x \rangle^*$$

PROBLEM–43

Which of the following wave functions are acceptable in quantum mechanics?

(a) $\sin x$ (b) $\tan x$

(c) $\operatorname{cosec} x$ (d) $\cos x + \sin x$

SOLUTION

The wave function (a) and (d) are acceptable wave functions while (b) and (c) are not because (b) tends to infinite at $x = \pi/2$ and (c) tends to infinite at $x = 0$.

PROBLEM–44

Write the Hamiltonian operator of a free particle moving in one direction under the influence of zero potential energy.

SOLUTION

In classical mechanics, Hamiltonian is given by the sum of K.E. and P.E. i.e.,

$$\hat{H} = \frac{1}{2} m x^2 + P.E.$$

But $$P.E. = 0$$

$\therefore$ $$\hat{H} = \frac{1}{2} m x^2 = \frac{1}{2m} p_x^2$$

where p_x is the linear momentum along x-axis.

But for linear operator the operator is $\frac{h}{i}\frac{d}{dx}$. Therefore,

$$\hat{H} = \frac{1}{2m}\left[\frac{h}{i}\frac{d}{dx}\right]$$

$$= \frac{1}{2m}\left[\frac{h}{i}\frac{d}{dx}\right]\left[\frac{h}{i}\frac{d}{dx}\right]^2$$

$$= \frac{h^2}{2m}\frac{d^2}{dx^2}$$

PROBLEM–45

Write a Hamiltonian for a freely moving particle and hence the corresponding Schrodinger wave equation.

SOLUTION

Suppose m be the mass of the particle which is moving with a velocity v_α in the direction of x. Then,

$$K.E. = T = \frac{1}{2} m v_\alpha^2 = \frac{p_x^2}{2m}$$

The classical Hamiltonian is given as follows :

$$H = K.E. + P.E.$$

$$\frac{p_x}{2m} + P.E. \tag{1}$$

But the potential energy is zero for freely moving particle. Therefore, Eq. (1) becomes as follows :

$$H = \frac{p_x^2}{2m} + 0$$

$$\frac{p_x^2}{2m} \tag{2}$$

But the operator for momentum is given as $\frac{h}{i}\frac{d}{dx}$. Therefore Eq. (2) becomes as follows :

$$\hat{H} = \frac{1}{2m}\left[\frac{h}{i}\frac{d}{dx}\right]\left[\frac{h}{i}\frac{d}{dx}\right]$$

$$= -\frac{h^2}{2m}\frac{d^2}{dx^2}$$

But we know, $\hat{H}\psi = E\psi$

$$-\frac{h^2}{2m}\frac{d^2\psi}{dx^2} = E\psi$$

or

$$\frac{d^2\psi}{dx^2} + \frac{h^2}{2m}\psi E = 0 \tag{3}$$

PROBLEM–46

Using the normalization wave function.

$$\psi_n(x) = (2/a)^{1/2} \sin\left(\frac{n\pi x}{a}\right) (n = 1, 2, 3, ...)$$

calculate the energy of a particle in a one-dimensional infinite box, of width equal to a.

SOLUTION

$$E_n \equiv E = \int_0^a \psi_n{*}(x)\, \hat{H}\, \psi_n(x)\, dx$$

$$= \int_0^a \left(\frac{2}{a}\right)^{1/2} \sin\left(\frac{n\pi x}{a}\right)\left[-\frac{h^2}{2m}\frac{d^2}{dx^2}\right]\left(\frac{2}{a}\right)^{1/2} \sin\left(\frac{n\pi x}{a}\right)dx$$

$$= \left(\frac{2}{a}\right)\left(-\frac{h^2}{2m}\right)\int_0^a \sin\left(\frac{n\pi x}{a}\right)\frac{d^2}{dx^2}\sin\left(\frac{n\pi x}{a}\right)dx$$

$$= \left(\frac{2}{a}\right)\left(-\frac{h^2}{2m}\right)\left(\frac{n\pi}{a}\right)^2(-1)\int_0^a \sin^2\left(\frac{n\pi x}{a}\right)dx \tag{1}$$

To evaluate the integral in Eq. (1)

let $$y = \frac{n\pi x}{a}.\ dy = \left(\frac{n\pi}{a}\right)dx;\ dx = \left(\frac{a}{n\pi}\right)dy$$

When $x = 0$, $y = 0$; when $x = a$, $y = n\pi$

Hence, $$\int_0^a \sin^2\left(\frac{n\pi x}{a}\right)dx = \left(\frac{a}{n\pi}\right)\int_0^{n\pi} \sin^2 y\, dy$$

$$= \left(\frac{a}{n\pi}\right)\left[\frac{y}{2} - \frac{\sin 2y}{4}\right]_0^{n\pi} = \left(\frac{a}{n\pi}\right)\left(\frac{n\pi}{2}\right) = \frac{a}{2}$$

Substituting in Eq. (1)

$$E_n = \left(\frac{2}{a}\right)\left(\frac{h^2}{2m}\right)\left(\frac{n\pi}{2}\right)^2\left(\frac{a}{2}\right)$$

$$= \frac{n^2\pi^2h^2}{2ma^2} = \frac{n^2h^2}{8ma^2}.$$

Note : the same result could be obtained directly from Eq. (1). Since the given wave function is normalized

$$\int \psi_n{}^*\psi_n\, dx = 1$$

or, $$\int\left(\frac{2}{a}\right)\sin^2\left(\frac{n\pi x}{a}\right)dx = 1$$

Substituting this value in Eq. (1)

$$E_n = \left(-\frac{h^2}{2m}\right)\left(\frac{n\pi}{2}\right)^2(-1) = \frac{n^2h^2}{8ma^2}.$$

PROBLEM–47

Calculate, from first principles, the ground-state energy of a particle in a square box, assuming that the potential is zero inside the box.

SOLUTION

$$\hat{H} = -\frac{h^2}{2m}\left(\frac{\partial^2}{\partial x^2} + \frac{\partial^2}{\partial y^2}\right)$$

The eigenfunctions of $\hat{H}$ are

$$\psi_{n_x, n_y} = \left(\frac{2}{a}\right)^{1/2} \sin\left(\frac{n_x \pi x}{x}\right)\left(\frac{2}{a}\right)^{1/2} \sin\left(\frac{n_y \pi y}{a}\right)$$

$$= \left(\frac{2}{a}\right) \sin\left(\frac{n_x \pi x}{a}\right) \sin\left(\frac{n_y \pi y}{a}\right)$$

For the ground state, $n_x = n_y = 1$

Hence,

$$\psi_{11} = \left(\frac{2}{a}\right) \sin\left(\frac{\pi x}{a}\right)\left(\frac{\pi y}{a}\right)$$

$$E_{n_x, n_y} = \int \psi^*{}_{n_x, n_y} \hat{H} \psi_{n_x, n_y}\, dx \equiv \langle \psi_{n_x, n_y} | \hat{H} | \psi_{n_x, n_y} \rangle$$

Thus,

$$E_{11} = \langle \psi_{11} | \hat{H} | \psi_{11} \rangle$$

$$< = \frac{2}{a} \sin\left(\frac{\pi y}{a}\right) \sin\left(\frac{\pi y}{a}\right) \left| -\frac{h^2}{2m}\left(\frac{\partial^2}{\partial x^2} + \frac{\partial^2}{\partial y^2}\right) \right| \frac{2}{a} \sin\left(\frac{\pi x}{a}\right) \sin\left(\frac{\pi y}{a}\right) >$$

$$= < \frac{2}{a} \sin\left(\frac{\pi x}{a}\right) \sin\left(\frac{\pi y}{a}\right) \left| \left(-\frac{h^2}{2m}(-2)\left(\frac{\pi}{a}\right)^2 \frac{2}{a} \sin\left(\frac{\pi x}{a}\right) \sin\left(\frac{\pi y}{a}\right) \right. >$$

$$= \frac{h^2}{m}\left(\frac{\pi}{a}\right)^2 < \frac{2}{a} \sin\left(\frac{\pi x}{a}\right) \sin\left(\frac{\pi y}{a}\right) \left| \frac{2}{a} \sin\left(\frac{\pi x}{a}\right) \sin\left(\frac{\pi y}{a}\right) \right. >$$

$$\frac{h^2 \pi^2}{ma^2} < \psi_{11} | \psi_{11} >$$

Sine ψ_{11} is normalized

$$< \psi_{11} | \psi_{11} > \equiv \int \psi_{11}^* \psi_{11}\, d\tau = 1$$

Hence,

$$E_{1,1} = \frac{h^2 \pi^2}{ma^2} = \frac{h^2}{4ma^2}$$

This result can be verified easily. For a particle in a square box the energy eigenvalues are given by

$$E_{n_x n_y} = \frac{h^2}{8ma^2}\left(n_x^2 + n_y^2\right)$$

with $n_x = n_y = 1$ this gives $h^2/4ma^2$) for $E_{1,1}$.

[Note : we have solved the above problem using Dirac's notation. This elegant notation is being frequently used in quantum mechanics. Here, the wave function ψ_n is written as the "ket vector" $|\psi_n>$; or, sometimes simply as $|n>$; while the complex conjugate wave function $\psi_n{}^*$ is written as the "bra vector" $<\psi_n|$; or, simply as $<n|$. Also,

$$\int \psi_n{}^*\psi_n d\tau \equiv <\psi_n | \psi_n> \equiv <n | n>$$

$$\int \psi_n{}^* \hat{H}\, \psi_n d\tau \equiv <\psi_n | \hat{H} | \psi_n> \equiv <n | \hat{H} | n> \equiv \hat{H}_{nn}$$

The quantity $\hat{H}_{nn'}$, is also called the matrix element of the Hamiltonian between the states $| n>$ and $| n'>$.]

PROBLEM–48

(a) Show that the wave functions for a particle in a one- dimensional box are orthonormal.

(b) Show that the wave functions for a particle in an infinite one-dimensional box

$$\psi_n(x) = N \sin\left(\frac{n\pi x}{a}\right) \quad (n = 0, 2, 4, 6, ...)$$

are not eigenfunctions of the linear momentum, p_x, but they can be written as combinations of eigenfunctions of p_x.

[Hint : (a)

$$\psi_n(x) = \left(\frac{2}{a}\right)^{1/2} \sin\left(\frac{n\pi x}{a} rig`t\right) \quad (n = 1, 2, 3, ...)$$

For orthonormality

$$\int_0^a \psi_n(x)\, \psi_m(x) dx = \delta_{nm} = \begin{cases} 1, & n = m \\ 0, & n \neq m \end{cases}$$

PROBLEM–49

Evaluate the normalization constant for a system of non- interacting particles moving along the x-axis if there are s_0 particles per unit length l. The value of l is chosen in such a way that $\cos l = 0$.

SOLUTION

The normalization is performed as follows :

$$\int_{x}^{x+l} |\psi(x)|^2 \, dx = s_0 l$$

i.e.

$$A^2 \int_{x}^{x+l} \sin^2\left(\frac{n\pi x}{l}\right) dx = \frac{A^2 l}{n\pi} \int_{n\pi x/l}^{n\pi(x+l)/l} \sin\beta \, d\beta = \frac{A^2 l}{2} = s_0 l$$

The normalized wave functions are thus $\psi(x) = \sqrt{2s_0}\sin(n\pi x/l)$. The energy values form a continuum.

PROBLEM–50

Obtain the expectation of x, x^2, p_x and p_x^2 for the particle in the box. Calculate the value of $(\Delta x)_{rms} \cdot (\Delta p_x)_{rms}$. In this in accordance with Heisenberg's uncertainty relation?

SOLUTION

Physical quantity x	x^2	p_x	p_x^2
Operator $X = x$	$X^2 = x^2$	$\frac{h}{2\pi i}\left(\frac{d}{dx}\right)$	$-\frac{h^2}{4\pi^2}\left(\frac{d^2}{dx^2}\right)$
Expectation value $\overline{x}$	$\overline{x^2}$	$\overline{p_x}$	$\overline{p_x^2}$

(i) Calculation of $\overline{x}$:

$$\overline{x} = \frac{2}{a}\int_0^a x\sin^2\left(\frac{n\pi x}{a}\right)dx = \frac{2a}{(n\pi)^2}\int_0^{n\pi} \beta\sin\beta \, d\beta$$

$$= \frac{2a}{(n\pi)^2}\left[\frac{\beta^2}{4} - \frac{\beta\sin 2\beta}{4} - \frac{\cos 2\beta}{8}\right]_0^{n\pi} - \frac{a}{2}$$

The average position of the particle is thus in the region where ψ^2 is highest.

(ii) Calculation of $\overline{x^2}$:

$$\overline{x^2} = \frac{2}{a}\int_0^a x^2\sin^2\left(\frac{n\pi x}{a}\right)dx = \frac{2a^2}{(n\pi)^2}\left[\frac{\beta^3}{6} - \left(\frac{\beta^2}{4} - \frac{1}{8}\right)\sin 2\beta - \beta\frac{\cos 2\beta}{4}\right]_0^{n\pi}$$

$$= \frac{a^2}{3}\left(1 - \frac{3}{2n^2\pi^2}\right)$$ The classical average of x^2 is :

$$\overline{x^2} = \int_0^a x^2 c\, dx \Big/ \int_0^a c\, dx = \frac{a^2}{3},$$ where $c = (\psi)^2$, a constant.

Note that the quantum mechanical $\overline{x^2}$ approaches the classical $\overline{x^2}$ for large values of n. This is an example of the *correspondence principle* of Bohr which states that the classical and quantum mechanical values converge when the system is in highly excited states.

(iii) Calculation of $\overline{p_x}$:

$$\overline{p_x} = \frac{2}{a}\int_0^a \sin\left(\frac{n\pi x}{a}\right)\left(\frac{h}{2\pi i}\frac{d}{dx}\right)\sin\left(\frac{n\pi x}{a}\right)dx = \frac{hn\pi}{a^2\pi i}\int_0^a \sin\left(\frac{n\pi x}{a}\right)\cos\left(\frac{n\pi x}{a}\right)dx$$

$$= 0$$

This corresponds with the classical picture of a particle moving back and forth (i.e. momentum $+p_x$ and $-p_x$) with constant velocity giving an average momentum of zero.

APPENDICES

Appendix A
Bibliography

Atkins, P.W., *Molecular Quantum Mechanics,* Clarendon Press, Oxford, 1970.

Barriol, Jean, *Elements of Quantum Mechanics with Chemical Applications,* Barnes Noble, New York, 1971.

Bohr, N., *Atomic Theory and Description of Nature,* Cambridge University Press, London, 1934.

Born, M., *Atomic Physics,* Blackie and Son, 1945.

Condon, E.U. and Shortley, G.H., *Theory of Atomic Spectra,* Cambridge University Press, 1963.

Cotton, F.A., *Chemical Applications of Group Theory,* Interscience, 1963.

Coulson, C.A., *Valence,* Oxford University Press, London, 1961.

Coulson, C.A., *Valence,* Second Edition, Oxford University Press, 1961.

Coulson, C.A., *Waves, A Mathematical Account of the Common Types of Wave Motion,* Oliver and Boyd, Edinburgh, 1961.

d'Abro, A., *The Rise of the New Physics,* Vol. II, Dover, New York, 1951.

Davis, J.C., *Advanced Physical Chemistry : Molecules, Structure, and Spectra,* Ronald Press, New York, 1965.

Davies, M., The electromagnetic spectrum in chemistry, *J. Chem. Educ.*, 31, 89, 1954.

de Broglie, L., *Matter and Light : The New Physics,* W.W. Norton, New York, 1939. (Reprinted by Dover, New York.)

Dirac, P.A.M., *Principles of Quantum Mechanics,* 4th ed., Oxford University Press, London, 1958.

Eisberg, R.M., *Fundamentals of Modern Physics,* John Wiley & Sons, New York, 1961.

Eyring, H., J. Walter, and G.E. Kimball, *Quantum Chemistry,* John Wiley & Sons, New York, 1944.

Eyring, H., Walter, J. and Kimbel,, G.E., *Quantum Chemistry,* Wiley, 1944.

Feynman, R.P., Leighton, R.B. and Sands, M., *The Feynman Lectures on Physics,* Vol. III, Addison-Wesley Publishing Co., Inc., 1965.

Gamow, G., *Mr. Tompkins in Wonderland,* Cambridge University Press, London, 1960.

Gamow, G., *Thirty Years That Shook Physics : The Story of Quantum Theory,* Doubleday & Co., New York, 1966.

Haliday, D., and R. Resnick, *Physics for Students of Science and Engineering,* Comb. Ed., John Wiley & Sons, New York, 1966.

Hameka, H.F., *Introduction to Quantum Theory,* Harper International, 1967.

Hanna, M.W., *Quantum Mechanics in Chemistry,* W.A. Benjamin, New York, 1965.

Harris, L., and A.L. Loeb, *Introduction to Wave Mechanics,* McGraw-Hill, New York, 1963.

Heisenberg, W., *The Physical Principles of Quantum Theory,* University of Chicago Press, Chicago, 1930. (Reprinted by Dover, New York.)

Heitler, W., *Elementary Wave Mechanics,* Oxford University Press, 1956.

Herzberg, G., *Spectra of Diatomic Molecules,* D. Van Nostrand, 1951.

Hochstrasser, R., *Molecular Aspects of Symmetry*, W.A. Benjamin, 1966.

Hoffmann, B., *The Strange Story of the Quantum*, 2nd ed., Dover, New York, 1959.

Hollas, J.M., *Symmetry in Molecules*, Chapman and Hall, 1972.

Kaufman, E.D., *Advanced Concepts in Physical Chemistry*, McGraw-Hill, New York, 1966.

Kauzmann, W., *Quantum Chemistry : An Introduction*, Academic Press, New York, 1957.

Kauzman, W., *Quantum Chemistry*, Academic Press, 1957.

Kotani, M., Ohno, K. and Kayama, K., Quantum mechanics of electronic structure of simple molecules, in *Hanbuch der Physik* (B and XXXVII/2 Molecules 11), Springer-Verlag, 1961.

Kramers, H.A., *Quantum Mechanics*, Dover, New York, 1964.

Lande, A., *Foundations of Quantum Theory*, Yale University Press, New Haven, 1955.

Linnett, J.W., *Wave Mechanics and Valency*, Methuen & Co., London, 1960.

Margenau, H., and G.M. Murphy, *The Mathematics of Physics and Chemistry*, D. Van Nostrand, New York, 1950.

Margenau, H. and Murphy, G.M., *Mathematics for Physics and Chemistry*, McGraw-Hill, 1956.

Murrell, J.N., Kettles, S.F.A. and Tedder, J.M., *Valence Theory*, Second Edition, John Wiley, 1970.

Murrell, J.N., *Theory of the Electronic Spectra of Organic Molecules*, Chapmann and Hall, 1970.

Pauling, L., and E.B. Wilson, *Introduction to Quantum Mechanics*, McGraw-Hill, New York, 1935.

Pauling, L. and Wilson, E.B., *Introduction to Quantum Mechanics*, McGraw-Hill, 1939.

Pilar, F.L., *Elementary Quantum Chemistry,* McGraw-Hill, New York, 1968.

Pillar, F.L., *Elements of Quantum Chemistry,* McGraw-Hill, 1969.

Pitzer, K.S., *Quantum Chemistry,* Prentice-Hall, Englewood Cliffs, N.J., 1953.

Rojansky, V., *Introductory Quantum Mechanics,* Prentice- Hall, Engle wood Cliffs, N.J., 1968.

Rojansky, V., *Introductory Quantum Mechanics,* Prentice- Hall, Inc., 1938.

Rose, M.E., *Elementary Theory of Angular Momentum,* Wiley, New York, 1961.

Schrodinger, E., *Science and Human Temperament,* W.W. Norton, New York, 1935.

Scherwin, C.W., *Introduction to Quantum Mechanics,* Holt, Rinehart and Winston, New York, 1959.

Shewall, J.R., On the formation of quantum mechanical operators, *Amer. J. Phys.* 27, 16, 1969.

Slater, J.C., *Quantum Theory of Atomic Structure,* Vol. I., McGraw-Hill, New York, 1960.

Slater, J.C., *Quantum Theory of Atomic Structure,* Vols., 1 & 2, Mc-Graw-Hill, 1960.

Slater. J.C., *Quantum Theory of Molecules and Solids,* Vol. 1, Mc-Graw-Hill, 1963.

Sokolnikoff, I.S., and E.S. Sokolnikoff, *Higher Mathematics for Engineers and Physicists,* McGraw-Hill, New York, 1934.

Steitwieser, A., *Molecular Orbital Theory for Organic* Chemists, Wiley, 1961.

Urch, D.S., *Orbitals and Symmetry*, Penguin Books Ltd., 1970.

Waldron, R.A., *Waves and Oscillations,* D. Van Nostrand, Princeton,. 1964.

Woodward, R.B. and Hoffmann, R., *Angew Chemie,* International Edition, 8, 781-853; 1969.

Appendix B
Greek Alphabet

Greek Alphabet

Alpha	A	α	Nu	N	ν
Beta	B	β	Xi	Ξ	ξ
Gamma	Γ	γ	Omicron	O	o
Delta	Δ	δ	Pi	Π	π
Epsilon	E	ε	Rho	P	ρ
Zeta	Z	ζ	Sigma	Σ	σ,s
Eta	H	η	Tau	T	τ
Theta	Θ	θ, ϑ	Upsilon	Υ	υ
Iota	I	ι	Phi	Φ	φ, ϕ
Kappa	K	κ	Chi	X	χ
Lambda	Λ	λ	Psi	Ψ	ψ
Mu	M	μ	Omega	Ω	ω

Appendix C
Hermitian Operators

In the specific case of the single particle in the one-dimensional box :

1. *Each of the eigenvalue energies is a real quantity.*
2. *The eigenfunctions of the Hamiltonian operator, each of which corresponds to a different eigenvalue energy, form an orthogonal set.*

These two very important conditions arise as a mathematical result of a property, called the *Hermitian property, which is possessed by all quantum-mechanical operators.* An operator $\hat{A}$ is said to be *Hermitian* if it obeys the relationship

$$\int \varphi_1{}^* \hat{A} \varphi_2 \, d\tau = \int \varphi_2{}^* \hat{A}^* \varphi_1{}^* \, d\tau \qquad \text{(c-1)}$$

where φ_1 and φ_2 are any two *well- behaved* functions whatever and where the integration is performed over the entire range of configurational space in which the two functions exist. The definition given by Eq. (c-1) also includes the specific case in which $\varphi_1 = \varphi_2$.

As a sample proof of the Hermitian nature of the quantum-mechanical operators given in Table 6-1 let's show that the linear momentum operator $\hat{p}_x$ is Hermitian over some finite range of one-dimensional space between the limits $x = a$ and $x = b$. That is, let's show that

$$\int_a^b \varphi_1{}^*\hat{p}_x\varphi_2\, dx = \int_a^b \varphi_2\hat{p}_x{}^*\varphi_1{}^*\, dx, \tag{c-2}$$

where φ_1 and φ_2 are functions of x.

Substituting of $h/i2\pi)(d/dx)$ for $\hat{p}_x$ and $(-h/i2\pi)(d/dx)$ for $\hat{p}_x{}^*$ into Eq. (c-2) yields

$$\frac{h}{i2\pi}\int_a^b \varphi_1{}^*\frac{d\varphi_2}{dx}\, dx = -\frac{h}{i2\pi}\int_a^b \varphi_2\frac{d\varphi_1{}^*}{dx}\, dx \tag{c-3}$$

which may be simplified and rearranged to

$$\int_{x=a}^{x=b} \varphi_1{}^*\, d\varphi_2 + \int_{x=a}^{x=b} \varphi_2\, d\varphi_1{}^* = 0. \tag{c-4}$$

But recall, for the general variables u and v, that

$$\int u\, dv + \int v\, du = \int d(uv)$$

so that Eq. (c-4) may be written as

$$\int_{x=a}^{x=b} d(\varphi_1{}^*\varphi_2) = \varphi_1{}^*\varphi_2\Big]_{x=a}^{x=b} = 0 \tag{c-5}$$

In evaluating the integral in Eq. (c-5) we must use the values of $\varphi_1{}^*\varphi_2$ only at the end points of the range. But we have insisted that all well-behaved functions have integrable squares, that is, that $\int \varphi_1{}^*\varphi_1\, dx$ and $\int \varphi_2{}^*\varphi_2\, dx$ each be finite, which means that φ_1 and φ_2 must each vanish at the end points of the range (for example, at 0 and L for the particle in the one- dimensional box). Thus

$$\varphi_1{}^*\varphi_2\Big]_0^0 = 0$$

or

$$0 = 0$$

and $\hat{p}_x$ is thus proved to be Hermitian.

Now let's show that *Hermitian operators always yield observables which are real numbers.* Even though i may appear as part of an operator or in the wave function itself, it is important that i should not appear in the allowed value of an observable if quantum mechanics is to have physical significance. We have shown that observables appear in quantum mechanics either as *eigenvalues A* according to the operator postulate (Postulate 2),

$$\hat{A}\psi = A\psi, \tag{c-6}$$

or as expectation or *average values* $\bar{A}$ according to the mean-value postulate (Postulate 5),

$$\bar{A} = \int \psi^* \hat{A}\psi \, d\tau \Big/ \int \psi^*\psi \, d\tau, \tag{c-7}$$

where the integrals are taken over all available space.

First, let's show that Hermitian operators always yield real eigenvalues for observables, as given by (c-7). The eigenvalue A_m generated as a result of the operation of $\hat{A}$ on an acceptable eigenfunction ψ_m is defined by

$$\hat{A}\psi_m = A_m\psi_m. \tag{c-8}$$

We may write the complex conjugate of Eq. (c-8) as

$$\hat{A}^*\psi_m^* = A_m^*\psi_m^*. \tag{c-9}$$

Multiplication of Eq. (c-8) by ψ_m^* followed by integration over all available space yields

$$\int \psi_m^*\hat{A}\psi_m \, d\tau = A_m \int \psi_m^*\psi_m \, d\tau. \tag{c-10}$$

In similar fashion, multiplication of Eq. (c-9) by ψ_m followed by integration over all available space yields

$$\int \psi_m \hat{A}^*\psi_m^* \, d\tau = A_m^* \int \psi_m^*\psi_m \, d\tau. \tag{c-11}$$

If $\hat{A}$ is a Hermitian operator, as defined by Eq. (c-1), the left-hand sides of Eqs. (c-10) and (c-11) are equal and combination of the two equations yields

$$A_m = A_m^*,$$

which is possible only when A_m is a real number. Thus, the physical significance of eigenvalues is assured by requiring that quantum-mechanical operators be Hermitian.

Now, let's also show that Hermitian operators always yield real expectation values for observables, as given by Eq. (c-7). Consider the evaluation of a certain observable A under conditions such that exact eigenvalues are not observed. The average value or expectation value

$\bar{A}_n$ for the observable in the state characterized by the wave function ψ_n is give by

$$\bar{A}_n = \int \psi_n{}^* \hat{A} \psi_n \, d\tau / \int \psi_n{}^* \psi_n \, d\tau, \tag{c-12}$$

where the integrations are performed over all available space. The complex conjugate of Eq. (c-12) is

$$\bar{A}_n{}^* = \int \psi_n{}^* \hat{A}^* \psi_n{}^* \, d\tau / \int \psi_n{}^* \psi_n{}^* \, d\tau, \tag{c-13}$$

But, if $\hat{A}$ is Hermitian,

$$\int \psi_n{}^* \, \hat{A} \, \psi_n \, d\tau = \int \psi n \, A^* \psi_n{}^* \, d\tau,$$

and the combination of Eqs. (c-12) and (c-13) then leads to

$$\bar{A}_n = \bar{A}_n{}^*,$$

which can be true only if $\bar{A}_n$ is a real quantity. Thus, the physical significance of the average value or expectation value of an observable is assured by requiring that all quantum-mechanical operators be Hermitian.

We have shown in the last section that the eigenfunctions for the single particle in the one-dimensional box form an orthogonal set. Now let's show under what specific conditions the Hermitian nature of the Hamiltonian operator $\hat{H}$ guarantees the orthogonality of wave functions. For convenience, we'll again restrict the argument to conservative systems. Recall that in any conservative system, the acceptable state functions for a particle are given, according to Postulate II, by the eigenfunctions of the time-independent Schrodinger equation :

$$\hat{H}\psi = E\psi.$$

Thus, for the two eigenfunctions ψ_m and ψ_n corresponding to the eigenvalue energies E_m and E_n we may write

$$H\psi_m = E_m\psi_m, \tag{c-14}$$

and

$$\hat{H}\psi_n = E_n\psi_n. \tag{c-15}$$

The complex conjugate of Eq. (c-14) is

$$\hat{H}^*\psi_m^* = E_m^*\psi_m^*. \tag{c-16}$$

But since $\hat{H}$ is Hermitian, E_m^* is equal to E_m because *subm* is real, and

$$\hat{H}^*\psi_m^* = E_m\psi_m^*. \tag{c-17}$$

Multiplication of Eq. (c-15) by ψ_m^* followed by integration over all available space yields

$$\int \psi_m^*\hat{H}\psi_n\, d\tau = E_n \int \psi_m^*\psi_n\, d\tau, \tag{c-18}$$

and multiplication of Eq. (c-17) by ψ_n followed by integration over all available space yields

$$\int \psi_n\hat{H}^*\psi_m^*\, d\tau = E_m \int \psi_m^*\psi_n\, d\tau. \tag{c-19}$$

Since $\hat{H}$ is a Hermitian operator, as defined by Eq. (c-1), the left-hand sides of Eqs. (c-18) and (c-19) are equal, so that the combination of the two equations leads to

$$(E_m - E_n) \int \psi_m^*\psi_n\, d\tau = 0. \tag{c-20}$$

It then follows that for two *different* eigenfunctions, for which E_m *is not equal to* E_n,

$$\int \psi_m^*\psi_n\, d\tau = 0 \ (m \neq n),$$

which is the condition for orthogonality. Thus, *eigenfunctions of the Hamiltonian operator (which are the state functions of the particle) are guaranteed to be orthogonal provided that they are associated with different energy eigenvalues, or with nondegenerate states.* The converse, however, is not true. That is, the fact that two different eigenfunctions of $\hat{H}$ are orthogonal does not mean that they must *necessarily* have different eigenvalues energies. We'll now show that even when the eigenvalues energies of two wave functions are *the same,* that is, even when states are degenerate, it is still possible to construct a suitable set of orthogonal functions by simple superposition. For example, consider the two *normalized* eigenfunctions ψ_m and ψ_n which have the common eigenvalue energy E :

$$\hat{H}\psi_m = E\psi_m,$$

and

$$\hat{H}\psi_n = E\psi_n.$$

If the two degenerate states are not orthogonal,

$$\int \psi_m{}^*\psi_n \, d\tau = b \neq 0, \tag{c-21}$$

where b is a finite constant. Let's, through appropriate superposition, construct the new function ψ'_n, where

$$\psi'_n = \psi_n - b\psi_m.$$

We may now show that ψ_m and ψ'_n are orthogonal; that is,

$$\int \psi_m{}^*\psi'_n \, d\tau = \int \psi_m{}^*\psi_n \, d\tau - b\int \psi_m{}^*\psi_m \, d\tau = b - b = 0.$$

In addition,

$$\hat{H}\psi'_n = \hat{H}(\psi_n - b\psi_m) = \hat{H}\psi_n - b\hat{H}\psi_m = E\psi_n - bE\psi_m = E\psi'_n$$

so that ψ'_n is also shown to be an eigenfunction of $\hat{H}$. We'll use a similar superposition in our later treatment of degenerate hybrid orbitals for the hydrogen atom.

Appendix D
Physical Constants and Conversion Factors

TABLE D-1
Defined Values and Equivalents

Unit	Abbreviations	Equivalent
Meter	m	1,650,763.73 wavelenghts *in vacuo* of the unperturbed transition $2p_{10}$–$5d_5$ in ^{86}Kr
Kilogram	kg	Mass of the international kilogram at Sevres, France
Second	sec	1/31, 556, 925.9747 of the tropical year at 12^h ET, 0 January 1900
Degree Kelvin	^oK	Defined in the Thermodynamic scale by assigning 273.16°K to the triple point of water (freezing point, 273.15°K = 0°C)
Unified atomic mass unit	amu	One-twelfth the mass of an atom of ^{12}C nuclide
Mole	mole	Amount of substance containing the same number of atoms (or other units) as 12 g of pure ^{12}C
Standard acceleration of free fall	g_n	9.806 65 m sec^{-2}, 980.665 cm sec^{-2}

(*contd...*)

Table D-1 contd...

Unit	Abbreviations	Equivalent
Normal atmospheric pressure	atm	101, 325 N m^{-2}, 1.1013,250 dyne cm^{-2}
Thermochemical calorie	cal_{th}	4.1840 J, 4.1840 × 10^7 erg
International steam table calorie	cal_{IT}	4.1868 J, 4.1868 × 10^7 erg
Liter	l	0.001000028 m^3, 1 000.028 cm^3 (recommended by CIPM, 1950)
Inch	in.	0.0254 m, 2.54 cm
Micron	μ	10^{-6} m, 10^{-4} cm
Angstrom	Å	10^{-10} m, 10^{-8} cm
Pound	lb	0.45359237 kg, 453.59237 g

N = newton = kg m sec–2, J = joule = newton-m = kg m2 sec–2

TABLE D-2
Adjusted Values of Constants

Constant	Symbol	Value	Estimated* error limit
Speed of light in vaccum	c	2.997925 × 10^{10} cm sec^{-1}	3
Elementary charge	e	4.80298 × 10^{-10} $cm^{3/2}$ $g^{1/2}$ sec^{-1} (esu)	20
Avogadro constant	N_A	6.02252 × 10^{23} $mole^{-1}$	28
Electron rest mass	m_e	9.1091 × 10^{-28} *g* 5.48597 × 10^{-4} *amu*	4 9
Proton rest mass	m_p	1.67252 × 10^{-24} *g* 1.00727663 × 10^0 *amu*	8 24
Neutron rest mass	m_n	1.67482 × 10^{-24} *g* 1.0086654 × 10^0 *amu*	8 13
Planck constant	h	6.6256 × 10^{-27} erg sec	5
Rydberg constant	R_∞ R_H	1.0973731 × 10^5 cm^{-1} 1.0967760 × 10^5 cm^{-1}	3 3
Bohr radius	a_0	5.29167 × 10^{-9} cm	7
Electron radius	r_e	2.81777 × 10^{-13} cm	11
Gas constant	R	8.3143 × 10^7 erg °K^{-1} $mole^{-1}$	12
Boltzmann constant	k	1.38054 × 10^{-16} erg °K^{-1}	18

* Based on three standard deviations.

TABLE D-3
Energy Conversion Factors

	erg	eV	J	*cal_{th}
1 erg =	1	6.2418×10^{11}	10–7	$2,3901 \times 10^{-8}$
1 ev =	1.6021×10^{-12}	1	1.6021×10^{-19}	3.8291×10^{-20}
1 J =	10^{7}	6.2418×10^{18}	1	0.23901
1 *cal_{th}	4.1840×10^{7}	2.6116×10^{19}	4.1840	1

* cal_{th} = thermochemical or defined calorie.

TABLE D-4
Ionization Potentials for the Elements

Atomic number	Symbol	Ionization Potential, eV/atom			
		I_1	I_2	I_3	I_4
1	H	13.595	—	—	—
2	He	24.581	54.403	—	—
3	Li	5.390	75.619	122.419	—
4	Be	9.320	18.206	153.850	217.7
5	B	8.296	25.149	37.920	259.3
6	C	11.256	24.376	47.871	64.5
7	N	14.53	29.593	47.426	77.4
8	O	13.614	35.108	54.886	77.4
9	F	17.418	34.98	62.646	87.1
10	Ne	21.559	41.07	63.5	97.0
11	Na	5.138	47.29	71.65	98.9
12	Mg	7.644	15.031	80.12	109.3
13	Al	5.984	18.823	28.44	120.0
14	Si	8.149	16.34	33.46	45.1
15	P	10.484	19.72	30.156	51.4
16	S	10.357	23.4	35.0	47.3
17	Cl	13.01	23.80	39.90	53.5
18	Ar	15.755	27.62	40.90	59.8
19	K	4.339	31.81	46	60.9
20	Ca	6.111	11.868	51.21	67
21	Sc	6.54	12.80	24.75	73.9
22	Ti	6.82	13.57	27.47	43.2

TABLE D-4
(continued....)

Atomic number	Symbol	Ionization Potential, eV/atom			
		I_1	I_2	I_3	I_4
23	V	6.74	14.65	29.31	48
24	Cr	6.764	16.49	30.95	50
25	Mn	7.432	15.636	33.69	~53
26	Fe	7.87	16.18	30.643	~56
27	Co	7.86	17.05	33.49	~53
28	Ni	7.633	18.15	35.16	—
29	Cu	7.724	20.29	36.83	—
30	Zn	9.391	17.96	39.70	—
31	Ga	6.00	20.51	30.70	64.2
32	Ge	7.88	15.93	34.21	45.7
33	As	9.81	18.63	28.34	50.1
34	Se	9.75	21.6	32	43
35	Br	13.996	24.56	35.9	47.3
36	Kr	13.996	24.56	36.9	—
37	Rb	4.176	27.5	40	—
38	Sr	5.692	11.027	25	57
39	Y	6.38	12.33	20.5	—
40	Zr	6.84	13.13	22.98	34.3
41	Nb	6.88	14.32	25.04	38.3
42	Mo	7.10	16.15	27.13	46.4
43	Tc	7.28	15.26	—	—
44	Ru	7.364	16.76	—	—
45	Rh	7.461	18.07	31.05	—
46	Pd	8.33	19.42	32.92	—
47	Ag	7.574	21.48	34.82	—
48	Cd	8.991	16.904	37.47	—
49	In	5.785	18.86	28.03	54.4
50	Sn	7.342	14.628	30.49	40.7
51	Sb	8.639	16.5	25.3	44.1
52	Te	9.01	18.6	31	38
53	I	10.454	19.09	—	—

TABLE D-4
(continued....)

Atomic number	Symbol	Ionization Potential, eV/atom			
		I_1	I_2	I_3	I_4
54	Xe	12.127	21.2	32.1	—
55	Cs	3.893	25.1	—	—
56	Ba	5.210	10.001	—	—
57	La	5.61	11.43	19.17	—
58	Ce	6.91	14.8	—	—
59	Pr	5.76	—	—	—
60	Nd	6.3	—	—	—
61	Pm	—	—	—	—
62	Sm	5.6	11.2	—	—
63	Eu	5.67	11.24	—	—
64	Gd	6.16	12	—	—
65	Tb	6.74	—	—	—
66	Dy	6.82	—	—	—
67	Ho	—	—	—	—
68	Er	—	—	—	—
69	Tm	—	—	—	—
70	Yb	6.22	12.10	—	—
71	Lu	6.15	14.7	—	—
72	Hf	7	14.9	—	—
73	Ta	7.88	16.2	—	—
74	W	7.98	17.7	—	—
75	Re	7.87	16.6	—	—
76	Os	8.7	17	—	—
77	Ir	9	—	—	—
78	Pt	9.0	18.56	—	—
79	Au	9.22	20.5	—	—
80	Hg	10.43	18.751	34.2	—
81	Tl	6.106	20.42	29.8	50.7
82	Pb	7.415	15.028	31.93	42.3
83	Bi	7.287	16.68	25.56	45.3
84	Po	8.43	—	—	—

TABLE D-4
(continued....)

Atomic number	Symbol	Ionization Potential, eV/atom			
		I_1	I_2	I_3	I_4
85	At	—	—	—	—
86	Rn	10.746	—	—	—
87	Fr	—	—	—	—
88	Rs	5.277	10.14	—	—